T0135736

TECHNISCHE UNIVERSITÄT MÜNCHEN
Zentrum Mathematik

Finite Element Error Analysis for PDE-constrained Optimal Control Problems: The Control Constrained Case Under Reduced Regularity

Dieter Sebastian Sirch

Vollständiger Abdruck der von der Fakultät für Mathematik der Technischen Universität München zur Erlangung des akademischen Grades eines

Doktors der Naturwissenschaften (Dr. rer. nat.)

genehmigten Dissertation.

Vorsitzender:		Univ.-Prof. Dr. Peter Rentrop
Prüfer der Dissertation:	1.	Univ.-Prof. Dr. Boris Vexler
	2.	Univ.-Prof. Dr. Thomas Apel
		Universität der Bundeswehr München

Die Dissertation wurde am 6.5.2010 bei der Technischen Universität München eingereicht und durch die Fakultät für Mathematik am 9.7.2010 angenommen.

Bibliografische Information der Deutschen Nationalbibliothek

Die Deutsche Nationalbibliothek verzeichnet diese Publikation in der
Deutschen Nationalbibliografie; detaillierte bibliografische Daten sind
im Internet über http://dnb.d-nb.de abrufbar.

ISBN 978-3-8325-2557-6

Logos Verlag Berlin GmbH
Comeniushof, Gubener Str. 47,
10243 Berlin
Tel.: +49 (0)30 42 85 10 90
Fax: +49 (0)30 42 85 10 92
INTERNET: http://www.logos-verlag.de

Abstract

Subject of this work is the analysis of numerical methods for the solution of optimal control problems governed by elliptic partial differential equations. Such problems arise, if one does not only want to simulate technical or physical processes but also wants to optimize them with the help of one or more influence variables. In many practical applications these influence variables, so called controls, cannot be chosen arbitrarily, but have to fulfill certain inequality constraints. The numerical treatment of such control constrained optimal control problems requires a discretization of the underlying infinite dimensional function spaces. To guarantee the quality of the numerical solution one has to estimate and to quantify the resulting approximation errors. In this thesis a priori error estimates for finite element discretizations are proved in case of corners or edges in the underlying domain and nonsmooth coefficients in the partial differential equation. These facts influence the regularity properties of the solution and require adapted meshes to get optimal convergence rates. Isotropic and anisotropic refinement strategies are given and error estimates in polygonal and prismatic domains are proved. The theoretical results are confirmed by numerical tests.

Zusammenfassung

Gegenstand der vorliegenden Arbeit ist die Untersuchung numerischer Methoden zur Lösung von Optimalsteuerproblemen mit elliptischen partiellen Differentialgleichungen als Nebenbedingung. Solche Aufgabenstellungen treten auf, wenn technische oder physikalische Prozesse nicht nur simuliert, sondern mit Hilfe einer oder mehrerer Einflussgrößen auch optimiert werden sollen. In vielen praktischen Anwendungen können diese Einflussgrößen, sogenannte Steuerungen, nicht beliebig gewählt werden, sondern unterliegen Ungleichungsbeschränkungen. Die rechentechnische Behandlung solcher steuerungsbeschränkter Optimalsteuerprobleme erfordert eine Diskretisierung der zugrundeliegenden unendlich dimensionalen Funktionenräume. Um die Qualität der numerischen Approximation sicherzustellen, müssen die durch die Diskretisierung entstehenden Abweichungen abgeschätzt und quantifiziert werden. In dieser Dissertation werden a priori Fehlerabschätzungen für Finite-Element-Diskretisierungen bewiesen, wenn das zugrundeliegende Rechengebiet Ecken oder Kanten aufweist oder die Zustandsgleichung nicht glatte Koeffizienten hat. Diese Umstände beeinflussen die Regularitätseigenschaften der Lösung und erfordern angepasste Netze um optimale Konvergenzraten zu erhalten. Es werden isotrope und anisotrope Verfeinerungsstrategien angegeben und Fehlerabschätzungen in polygonalen und prismatischen Gebieten bewiesen. Die theoretischen Resultate werden jeweils durch numerische Tests bestätigt.

Vorwort

Die vorliegende Dissertation entstand während meiner Tätigkeit als wissenschaftlicher Mitarbeiter am Institut für Mathematik und Bauinformatik der Fakultät für Bauingenieur- und Vermessungswesen an der Universität der Bundeswehr München. Mein Dank richtet sich zuallererst an Prof. Dr. Thomas Apel für seine große Unterstützung und das mir entgegengebrachte Vertrauen. Er hatte stets ein offenes Ohr für Probleme, investierte viele Stunden in zahlreiche Diskussionen und gab immer wieder interessante Anregungen. Thomas, vielen Dank dafür. Danken möchte ich auch meinen Kollegen am Institut für die gute Arbeitsatmosphäre und das angenehme gesellschaftliche Miteinander innerhalb und außerhalb der Universität.

Mein besonderer Dank gilt Prof. Dr. Boris Vexler, der bereit war, die Betreuung der Arbeit an der TU München zu übernehmen. Ein herzliches Dankeschön richtet sich auch an Prof. Dr. Serge Nicaise, der mir zwei Forschungsaufenthalte an der Universität in Valenciennes ermöglichte und mir immer mit Rat und Tat zur Seite stand. Für die gute Zusammenarbeit möchte ich mich auch bei Prof. Dr. Arnd Rösch bedanken. Prof. Dr. Gunar Matthies danke ich für die Bereitstellung des Softwarepakets MoonMD und seine stete Hilfsbereitschaft bei auftretenden Problemen.

Zu guter Letzt möchte ich meine große Dankbarkeit gegenüber meiner Familie aussprechen. Meine Eltern haben mich über die vielen Jahre von Schule, Studium und Promotionszeit hinweg stets in vielfacher Weise unterstützt und damit einen ganz besonderen Anteil am Gelingen dieser Arbeit. Mein Dank richtet sich auch an meine Freundin Diana, die mir immer Rückhalt und Aufmunterung für meine Arbeit gab.

Contents

CHAPTER **1**

Introduction

1.1 Motivation

The modelling and numerical simulation of complex systems play an important role in many industrial, medical and economical applications. Very often, such systems can mathematically be described by partial differential equations (PDEs). Here, one can think for example of heat flow in materials or human tissues, aerodynamic properties of airplanes or determination of option prices in finance. In the last decades the development of efficient numerical methods to solve PDEs gave people together with the rising computing power the opportunity to simulate complex systems. Today this is done very successfully in many areas. But in most applications mathematical modelling and numerical simulation are only the first steps. People are rather interested in optimization or optimal control of the simulated processes. Examples are optimal control of the hydration of concrete [4], optimal control of glass cooling [73] or optimal placement of a probe in cancer therapy [2]. In optimal control problems the optimization variable is typically split in two parts, namely in a control variable and in a state variable, which is influenced by the control variable. In our case control and state are coupled by a partial differential equation such that a given control determines a unique state by the solution of this PDE. The aim is to find a control such that the state minimizes a certain quantity. Here, one can think for example of the temperature in a furnace that has to be controlled such that the hot melt of glass inside is cooled as close as possible along an optimal temperature curve to avoid cracks or to affect the optical quality of the resulting product. Since one cannot adjust the temperature in a furnace arbitrarily, one has some constraints on the control. In practise, often constraints on the state or the gradient of the state occur, e. g., the glass temperature should not exceed a certain level and must not be cooled too fast. In this thesis we consider only problems with control constraints. Additional difficulties occur if one has to deal with nonconvex geometries, e.g., in the modelling of a new concrete wall on a bottom plate [4], or with nonsmooth coefficients in the underlying PDE as it can happen, e.g., when modelling the heat distribution in a solid that consists of different materials.

Such optimal control problems can abstractly be written as

$$\min_{y \in Y, u \in U} J(y, u) \quad \text{subject to} \quad e(y, u) = 0, \ u \in \mathcal{C}, \tag{1.1}$$

where Y and U are Banach spaces, y is the state variable and u the control variable. The term $e(y, u) = 0$ denotes a PDE and the set $\mathcal{C}$ is a closed convex subset of U. In general one is not able to give an analytical solution of problem (1.1) and has to rely on numerical methods. In order to obtain stable and accurate numerical results one has to explore and utilize the specific mathematical structure behind and develop intelligent discretization strategies. A main ingredient in the analysis are error estimates. They make it possible to find reasonable discretizations and to validate numerical results. This thesis contributes to this topic and gives a priori error estimates for control constrained linear-quadratic optimal control problems governed by elliptic PDEs when additional singularities occur caused by corners or edges in the domain or by only piecewise smooth coefficients in the state equation.

1.2 State equation

We discretize the optimal control problem by a finite element method. Since the finite element error in the state equation plays an important role in the error analysis of the optimal control problem, we first concentrate on elliptic boundary value problems in nonsmooth domains or with only piecewise smooth coefficients. Before one can start with the error analysis one has to figure out the regularity properties of the solution. The literature on this topic is vast, such that is impossible to give an exhaustive overview here. Let us mention at least the fundamental paper of Kondrat'ev [77] and the monographs of Grisvard [62, 63], Kufner and Sändig [80] and Dauge [45]. Furthermore the books of Nazarov and Plamenvsky [99] and Kozlov, Maz'ya and Rossmann [78, 79] summarize the research of the authors in this field over many years. It turns out that the solution of such problems can be characterized by a so called singularity exponent λ. In detail, it is contained in the Sobolev-Slobodetski space $H^s(\Omega)$ with $s < 1 + \lambda$ provided the right-hand side is smooth enough. Consequently, one can expect for a finite element method with polynomial shape functions of order k on quasiuniform meshes only a reduced convergence order in $H^1(\Omega)$ or $L^2(\Omega)$, namely convergence rates of λ or 2λ in the discretization parameter h. The optimal order can only be reached if the solution is contained in $H^{k+1}(\Omega)$.

To improve the convergence properties under reduced regularity researchers started to develop specially adapted methods. Oganesyan and Rukhovec [103], Raugel [110] and Babuška, Kellog and Pitkäranta [22] investigated a priori local mesh grading techniques in the two dimensional case and got the same rates in $L^2(\Omega)$ and $H^1(\Omega)$ as in the regular case. We are also interested in estimates of the pointwise error. Many results on this topic were published in the 1970's, see, e.g., [58, 98, 117, 118, 122]. Scott proved in [122] a convergence rate of $h^2 |\ln h|$ for an elliptic equation and Neumann boundary conditions. This result is valid if the solution is in $W^{2,\infty}(\Omega)$ and if the mesh is quasi-uniform. Frehse and Rannacher considered in [58] the Dirichlet problem for a more general elliptic operator in domains Ω with $\partial\Omega \in C^{2,\alpha}$ and for a discretization with quasi-uniform meshes. For solutions in $W^{2,\infty}(\Omega)$ they got the convergence rate $h^2 |\ln h|$, for a right-hand side from $L^\infty(\Omega)$ they proved the approximation order $h^2 |\ln h|^2$. Since we like to consider domains with corners, the boundary is not in $C^{2,\alpha}$ and the state is in general not in $W^{2,\infty}(\Omega)$. So these results are not applicable. In [117] Schatz and Wahlbin derived pointwise estimates for the Poisson equation in domains with corners. In [118] they specified a refinement rule for the mesh that allows to prove a convergence rate of $h^{2-\epsilon}$ using piecewise linear ansatz functions. The drawback of this result is the fact, that the error constant is not separated from a norm of the right-hand side of the boundary value problem. Especially it is not clear, what regularity has to be assumed for the right-hand side, since Schatz and Wahlbin only demand a "smooth" right-hand side. We would like to emphasize that in the case of optimal control the right-hand side is the unknown control and therefore one cannot assume arbitrary smoothness.

For L^2- and H^1-error estimates Apel and Heinrich [8], Apel, Sändig and Whiteman [16] and Lubuma and Nicaise [86] extended the mesh grading idea to the three-dimensional case. They used piecewise linear approximations on *isotropic meshes*, i.e., the aspect ratio of the finite elements was bounded. But it was observed that this technique leads to overrefinement near edges. In order to avoid this overrefinement, *anisotropic meshes* in the neighborhood of the edges were used in [7, 10, 6, 12, 13]. Anisotropic finite elements are more general than shape-regular elements; they are characterized by three size parameters $h_{i,T}$, $i = 1, 2, 3$, which may have different asymptotics. The anisotropic mesh grading is described by a relationship between the size parameters of each element and its distance from an edge. By estimating the approximation error of the standard nodal interpolation operator and using the projection property of the finite element method, it is shown in [7, 10] that the finite element solution using a linear ansatz space converges like $O(h)$ in $H^1(\Omega)$ to the solution as long as the right-hand side is in $L^p(\Omega)$, $p > 2$. Here, h is as usual the maximum diameter of all elements. The main drawback of this estimate is, that the case $p = 2$ cannot be treated in this way and an L^2-estimate of the finite element error cannot be obtained. The reason for this can be found in fact that by the use of the standard Lagrangian (nodal) interpolant the local interpolation error does not converge in $H^1(T)$ with order 1 in h, if T is an anisotropic element and the solution is only in $H^2(\Omega)$, see [7]. As remedy Apel introduced in [6] suitable Scott-Zhang type quasi-interpolants. The disadvantage of these modified operators is that they preserve Dirichlet boundary conditions on parts of the boundary only and that the analysis is made for meshes with certain structure only. Nevertheless, this allowed to prove second order convergence in $L^2(\Omega)$ on appropriately graded meshes for a mixed boundary value problem in a prismatic domain with reentrant edge, where the part with the Dirichlet conditions was chosen such that one of these modified Scott-Zhang operators could preserve them. One main ingredient of the proof was also the description of the regularity of the solution in certain weighted Sobolev spaces. A non-conforming approximation for the Poisson equation with pure Dirichlet boundary conditions in a prismatic domain with reentrant edge was considered in [12]. The authors proved second order convergence in $L^2(\Omega)$ for a finite element approximation in the lower order Crouzeix-Raviart finite element space on appropriately graded anisotropic meshes. These results were extended in [13] to the Stokes problem.

Beyond the described a priori local mesh grading technique there exists a couple of other methods to treat singularities. Let us mention here at least the singular function method. The basic idea is to augment the ansatz space by certain functions that describe the occurring singularities. For a detailed description we refer to [29, 54, 123] and for the three-dimensional case also to [27, 87]. A similar approach is propagated in [23, 31] where the singular part of the solution is calculated explicitly.

1.3 Optimal control problems

Let us come back to the optimal control problem. The a priori error analysis for optimal control problems started with the papers of Falk [57] and Gevici [60]. They followed the classical approach of discretizing both, state and control, with piecewise polynomial finite elements. Particularly, they investigated the case of elliptic state equations and pure control constraints and considered piecewise constant approximations of the control. Malanowski discussed in [88] piecewise constant and piecewise linear approximations in space for a parabolic problem. In the last years optimal control of PDEs became popular again and researchers restarted to investigate numerical schemes for such problems. Arada, Casas and Tröltzsch [20] and Casas, Mateos and Tröltzsch [38] considered semilinear equations with piecewise constant approximations of the control and got a convergence rate of 1 in the discretization parameter h in $L^2(\Omega)$. Rösch investigated in [111] an abstract optimal control problem and proved under certain assumptions a convergence rate of $3/2$ in $L^2(\Omega)$ for piecewise linear approximations of the control. Casas and Tröltzsch considered in [39] also piecewise linear approximations of the control and proved first order convergence for several elliptic problems in general situations. This result was improved to superlinear convergence in a paper by Casas [36]. Arada, Casas and Tröltzsch proved in [20] also L^∞-error estimates. They got a convergence rate of 1 in 2D and of $1/2$ in 3D. Meyer and Rösch gave pointwise error estimates for piecewise linear approximations in [96] and proved first order convergence. All the mentioned papers assume quasi-uniform meshes and sufficiently smooth solutions.

A different approach is proposed by Hinze [71]. In his variational discretization concept the space of admissible controls is not discretized. Instead, the first order optimality condition and the discretization of the state and the adjoint state are utilized to derive an approximate control. It was proved that the discretization error of the control is bounded by finite element errors, such that this error behaves for piecewise linear approximations of state and adjoint state like $O(h^2)$ and $O(h^2 |\ln h|)$ in the L^2- and the L^∞-norm, respectively, where full regularity of the state and the adjoint state was assumed. The same L^2-estimate was proved in [19] under reduced regularity assumptions for appropriately graded, isotropic meshes.

Another discretization concept was introduced by Meyer and Rösch [95]. The space of admissible controls is discretized by piecewise constant functions. The final approximation is computed in a post-processing step that consists of a projection of the scaled approximate adjoint state in the set of admissible controls. The authors proved second order convergence for plane, convex domains under the assumption of full regularity in state and adjoint state. Apel, Rösch and Winkler proved in [15] the same result for non-convex plane domains with the use of local mesh grading. The article of Apel and Winkler [19] extended the results to general three-dimensional domains, where state and adjoint state may not admit the full regularity. They counteracted the impact of singularities, which are caused by reentrant corners and edges, by isotropic, graded meshes and proved a convergence rate of 2. In order to avoid overrefinement near edges Winkler considered in [127] an anisotropic discretization for a linear-quadratic optimal control problem with a special type of mixed boundary conditions in the state equation and proved also second order convergence. This result was proved for meshes with grading parameter $\mu < \min\{\lambda, 5/9 + \lambda/3\}$, where λ is the singularity exponent mentioned above. A detailed definition of these quantities can be found in Chapter 2. This is a stronger condition than actually necessary to get optimal convergence for the state equation itself, where $\mu < \lambda$ is enough, see [6]. In [40] Chen considered a mixed formulation of the elliptic state equation and derived superconvergence results for the postprocessing approach for Raviart-Thomas finite element discretizations on rectangular domains.

Rösch and Vexler applied in [112] the post-processing technique to a linear-quadratic optimal control problem with the Stokes equations as state equations. They achieved second order convergence provided that no singularities occur such that the velocity field is in $H^2(\Omega) \cap W^{1,\infty}(\Omega)$. Therefore they restricted theirselves to polygonal, convex domains $\Omega \subset \mathbb{R}^d$, $d = 2, 3$, and assume in the case $d = 3$ that the edge openings of the domain Ω are smaller than $2\pi/3$. Casas et al. considered in [37] locally constrained optimal control problems with the steady-state Navier-Stokes equations in smooth domains. We should also mention that several articles were published for the optimal control of the Stokes and Navier-Stokes equation without control constraints, see e.g. [30, 50, 67, 68].

Although this thesis concentrates on the control-constrained case let us also mention a couple of papers that are devoted to the a priori error analysis of problems with constraints on the state [41, 51, 52, 53, 72, 93, 94] or the gradient of the state [49, 65, 105]. Beyond the a priori analysis in the last years several researchers contributed to the a posteriori analysis of optimal control problems. We point to the papers [59, 69, 70, 83, 85, 125] for the control-constrained case and [28, 66, 74] for problems with state constraints.

1.4 Outline

The outline of the thesis is as follows. In Chapter 2 we repeat some basic facts from functional analysis, define the function spaces we need in our analysis and give corresponding embedding results. Furthermore, we introduce the graded meshes that are used for domains with singular corners or edges.

Chapter 3 is devoted to interpolation. Particularly we treat nonsmooth functions on anisotropic finite elements. We extend results for a quasi-interpolation operator from [6] such that also pure Dirichlet and pure Neumann problems can be treated. The specific difficulty with the Dirichlet problem is that the relevant quasi-interpolation operator E_h defined in [6] does not preserve the boundary conditions on the whole boundary but only on a part of it. To solve this problem we define a modification of this operator and estimate the additionally occurring error term. The Neumann problem was not satisfactorily treated in [6] since its solution has to be described in other weighted Sobolev spaces than the Dirichlet and the mixed problems. The reason is that in case of Neumann conditions on both faces joining the "singular edge" the solution may not vanish along the edge. In view of this, the proof of the global interpolation error estimate in [6, Theorem 14] is wrong for the Neumann case. To overcome this problem we prove local estimates in the corresponding weighted Sobolev spaces. This builds up the basis for the estimate of the global error.

Chapter 4 contains a couple of finite element error estimates for boundary value problems that serve as state equation for optimal control problems in Chapter 5. In the first section of Chapter 4 we prove an L^∞-error estimate for scalar elliptic problems with Hölder continuous right-hand sides in domains with reentrant corner on graded meshes. The novelty of this result is the fact that the error constant in this estimate is separated from the norm of the right-hand side of the boundary value problem, see Theorem 4.4. Therefore this result is applicable in the context of optimal control, where the right-hand side is the unknown control such that one cannot assume arbitrary smoothness as it is done e.g. in [118]. Section 4.2 utilizes the results of Chapter 3 to prove global interpolation error estimates in $L^2(\Omega)$ for an elliptic equation with pure Dirichlet or Neumann boundary conditions. This gives us straightforwardly an estimate of the corresponding finite element errors. Let us also mention, that we prove in this section the boundedness of $r^\beta \nabla y$ for $\beta > 1 - \lambda$, where r is the distance to the edge and y the solution of the boundary value problem, see Lemma 4.20. This is an improvement of the condition $\beta > 4/3 - \lambda$ as proved in [127] and allows in contrast to that thesis to keep the same grading condition for boundary value problem and optimal control problem. In Section 4.3 problems with discontinuous diffusion coefficients are considered and finite element error estimates are given. Section 4.4 treats the Stokes equations with Dirichlet boundary conditions. We first prove error estimates under some reasonable assumptions. Afterwards we show that these assumptions are fulfilled for the Stokes problem in a two-dimensional domain with reentrant corner and a three-dimensional domain with reentrant edge. In the first case we consider several commonly used conforming element pairs as well as the lower order Crouzeix-Raviart element, both on isotropic graded meshes. For the three-dimensional setting we investigate a discretization on anisotropic graded meshes with Crouzeix-Raviart elements and prove in addition to [13] also error estimates in $L^2(\Omega)$.

In Chapter 5 we turn to optimal control problems. First, we define in Section 5.1 a general linear-quadratic optimal control problem and prove L^2-error estimates for all three variables under certain assumptions on the discretization for both, the variational discrete approach introduced in [71] and the post-processing approach introduced in [95]. For the post-processing approach this general formulation is new. To the best of the author's knowledge this is the first time that also non-conforming discretizations are allowed.

We begin the consideration of particular examples in Section 5.2 with problems with scalar elliptic state equation. We prove that in polygonal domains the approximation error in the L^∞-norm behaves like $O(h^2 |\ln h|^{3/2})$. Notice that this estimate is new for the postprocessing approach even in the case of convex domains. For the variational discrete approach there is an estimate in [71] which depends on the finite element error of the adjoint equation in $L^\infty(\Omega)$. As example the suboptimal rates of $h^{2-d/2}$ in space dimension d are pointed out as well as the optimal rate of $h^2|\log h|$ for solutions in $W^{2,\infty}(\Omega)$. But improved estimates for domains with obtuse angles are not considered in that paper. Next, we check the assumptions of Section 5.1 for optimal control problems in prismatic domains and discretizations on anisotropic finite element meshes. In this way, we extend the L^2-error estimates of [127] to the case of pure Dirichlet and pure Neumann boundary conditions. A challenge in case of Neumann boundary conditions are the different regularity properties compared to the Dirichlet case. This requires some significant changes in the proofs of [127]. We further weaken the mesh grading condition given in that thesis to $\mu < \lambda$, what is the same as one has to demand to get optimal convergence in the state and adjoint state equation. We have to pay with slightly more regularity in the desired state, which has to be Hölder continuous and not only bounded. As a byproduct we can also weaken the grading condition for isotropic refinement given in [19]. We finish Section 5.2 with error estimates for an optimal control problem with a state equation with nonsmooth coefficients

We continue in Section 5.3 with an example where a linear-quadratic functional has to be minimized with respect to the Stokes equations. We consider a nonconvex prismatic domain which is discretized by an anisotropic graded mesh and approximate the velocity in the Crouzeix-Raviart finite element space. For the check of the assumptions of Section 5.1 we do not only have to deal with the more complicate structure of the Stokes equations but also with missing regularity in edge direction. In contrast to the Poisson equation, we do not have additional regularity of the solution and its derivatives in edge direction in $L^p(\Omega)$ for general p. In the case of the Stokes equations such results are only available for $p = 2$, see [13]. This fact prevents a componentwise application of the arguments which were used in case of the Poisson equation and makes new ideas necessary. Our last example concerns a two-dimensional setting, where the domain has a reentrant corner. We prove that our general assumptions are satisfied for a couple of element pairs as long as one uses a mesh that is tailored to the corner singularity. All the results in Chapter 5 are illustrated by numerical examples.

In the last chapter we conclude our results and give an outlook on future work.

CHAPTER 2

Preliminaries

2.1 Basic facts from functional analysis

In this section we recall some basics from functional analysis. Details can be found in any book on linear functional analysis, e.g., [1, 126].

Definition 2.1. Let X be a real vector space.

(1) A mapping $\|\cdot\| : X \to \mathbb{R}$ is called *norm* on X, if for $x, y \in X$ and $\alpha \in \mathbb{R}$ the conditions

    (i) $\|x\| \geq 0$ and $\|x\| = 0 \Leftrightarrow x = 0$,

    (ii) $\|\alpha x\| = |\alpha| \|x\|$,

    (iii) $\|x + y\| \leq \|x\| + \|y\|$

hold.

(2) The pair $(X, \|\cdot\|)$ is called *normed space*.

(3) A normed space is called *complete*, if any Cauchy sequence (x_n) has a limit, i.e., if $\lim_{m,n \to \infty} \|x_m - x_n\| = 0$ implies the existence of $x \in X$ with $\lim_{n \to \infty} \|x_n - x\| = 0$.

(4) A normed space, that is complete, is called *Banach space*.

Definition 2.2. Let H be a real vector space.

(1) A mapping $(\cdot, \cdot) : H \to \mathbb{R}$ is called *inner product* on H, if for $x, y, z \in H$ and $\alpha \in \mathbb{R}$ the conditions

    (i) $(x, y) = (y, x)$,

    (ii) $(\alpha x, y) = \alpha(x, y)$,

(iii) $(x + y, z) = (x, z) + (y, z)$,

(iv) $(x, x) \geq 0$ and $(x, x) = 0 \Leftrightarrow x = 0$

are satisfied.

(2) The pair $(H, (\cdot, \cdot))$ is called *Pre-Hilbert space*.

(3) A Pre-Hilbert space is called *Hilbert space*, if it is complete under its associated norm $\|x\| := \sqrt{(x, x)}$.

Theorem 2.3. *In a Pre-Hilbert space H the* Cauchy-Schwarz inequality

$$|(x, y)| \leq \|x\| \, \|y\| \quad \forall x, y \in H$$

holds.

Definition 2.4. Let X, Y be normed real vector spaces with norms $\| \cdot \|_X$, $\| \cdot \|_Y$. A mapping $A : X \to Y$ is called *linear operator* if it satisfies

$$A(\lambda x_1 + \mu x_2) = \lambda A x_1 + \mu A x_2 \quad \forall x_1, x_2 \in X, \ \lambda, \mu \in \mathbb{R}.$$

The space of all linear operators $A : X \to Y$ that are bounded in the sense that

$$\|A\|_{X \to Y} := \sup_{\|x\|_X = 1} \|Ax\|_Y < \infty$$

is denoted by $\mathcal{L}(X, Y)$.

Definition 2.5. Let X be a Banach space. The space $X^* := \mathcal{L}(X, \mathbb{R})$ of linear functionals on X is called *dual space* of X. We use the notation

$$\langle x^*, x \rangle_{X^*, X} := x^*(x).$$

The term $\langle \cdot, \cdot \rangle_{X^*, X}$ is called *dual pairing* of X^* and X.

Definition 2.6. Let X, Y be Banach spaces and $A \in \mathcal{L}(X, Y)$. Then the operator $A^* \in \mathcal{L}(Y^*, X^*)$ defined by

$$\langle A^* y^*, x \rangle_{X^*, X} = \langle y^*, Ax \rangle_{Y^*, Y} \quad \forall y^* \in Y^*, x \in X$$

is called *dual operator* of A.

Definition 2.7. Let X and Y be two normed spaces. X is said to be *embedded* into Y, written $X \hookrightarrow Y$, if there is a constant c such that for all $x \in X$ possibly after modification on a set of measure zero $x \in Y$ and $\|x\|_Y \leq c\|x\|_X$, i.e., the embedding operator $T : x \in X \mapsto x \in Y$ is bounded.

X is said to be *compactly embedded* into Y, written $X \overset{c}{\hookrightarrow} Y$, if $X \hookrightarrow Y$ and every sequence (x_n) which is bounded in X has a subsequence which converges in Y, i.e. the embedding operator $T : x \in X \mapsto x \in Y$ is compact.

2.2 Function spaces

This section covers function spaces which are needed to classify solutions of boundary value problems and optimal control problems. Beyond the Lebesgue spaces $L^p(G)$ and the classical Sobolev spaces $W^{k,p}(G)$ we shall need some weighted spaces $V_\beta^{k,p}(G)$ and $W_\beta^{k,p}(G)$, which are tailored to the smoothness properties of solutions of boundary value problems in nonconvex domains G or with discontinuous coefficients. Moreover, we introduce the space of Hölder continuous functions $C^{0,\sigma}(\bar{G})$.

Throughout this section let $G \subset \mathbb{R}^d$, $d = 2, 3$, be an open, bounded domain with Lipschitz boundary ∂G. The closure of the domain G is denoted by $\bar{G}$. Further, we denote by $\mathcal{D}'(G)$ the space of distributions.

Definition 2.8. We define for $p \in [1, \infty)$ the *Lebesgue space* $L^p(\Omega)$ by

$$L^p(G) := \left\{ v : G \to \mathbb{R} \text{ Lebesgue measurable} : \|v\|_{L^p(G)} := \left(\int_G |u(x)|^p \right)^{1/p} < \infty \right\}.$$

Moreover, we define

$$L^\infty(G) := \left\{ v : G \to \mathbb{R} \text{ Lebesgue measurable} : \|v\|_{L^\infty(G)} := \operatorname{ess\,sup}_{x \in G} |u(x)| < \infty \right\}.$$

Definition 2.9. We define for $k \in \mathbb{N}$, $p \in [1, \infty)$ the *classical Sobolev spaces* $W^{k,p}(G)$ by

$$W^{k,p}(G) := \left\{ v \in \mathcal{D}'(G), \|v\|_{W^{k,p}(G)} := \left(\sum_{|\alpha| \leq k} \int_G |D^\alpha v|^p \right)^{1/p} < \infty \right\}$$

Furthermore, we introduce the *seminorm*

$$|v|_{W^{k,p}(G)} := \left(\sum_{|\alpha| = k} \int_G |D^\alpha v|^p \right)^{1/p}.$$

We also use the abbreviation $H^k(G) := W^{k,2}(G)$ and define $W^{0,p}(G) := L^p(G)$. Finally, we define

$$W^{k,\infty}(G) := \left\{ v \in \mathcal{D}'(G), \|v\|_{W^{k,\infty}(G)} := \max_{|\alpha| \leq k} \|D^\alpha v\|_{L^\infty(G)} < \infty \right\}.$$

Definition 2.10. With polar/cylindrical coordinates $x_1 = r \cos\varphi$, $x_2 = r \sin\varphi$, we define for $k \in \mathbb{N}_0$, $p \in [1, \infty)$ and $\beta \in \mathbb{R}$ the *weighted Sobolev spaces*

$$V_\beta^{k,p}(G) := \left\{ v \in \mathcal{D}'(G) : \|v\|_{V_\beta^{k,p}(G)} := \left(\sum_{|\alpha| \leq k} \int_G |r^{\beta - k + |\alpha|} D^\alpha v|^p \right)^{1/p} < \infty \right\},$$

$$W_\beta^{k,p}(G) := \left\{ v \in \mathcal{D}'(G) : \|v\|_{W_\beta^{k,p}(G)} := \left(\sum_{|\alpha| \leq k} \int_G |r^\beta D^\alpha v|^p \right)^{1/p} < \infty \right\}.$$

The corresponding *seminorms* are defined as

$$|v|_{V_\beta^{k,p}(G)} := \left(\sum_{|\alpha|=k} \int_G |r^{\beta-k+|\alpha|} D^\alpha v|^p \right)^{1/p} \quad \text{and} \quad |v|_{W_\beta^{k,p}(G)} := \left(\sum_{|\alpha|=k} \int_G |r^\beta D^\alpha v|^p \right)^{1/p}.$$

Moreover, we introduce

$$V_\beta^{k,\infty}(G) := \left\{ v \in \mathcal{D}'(G) : \|v\|_{V_\beta^{k,\infty}(G)} := \max_{|\alpha|\leq k} \|r^{\beta-k+|\alpha|} D^\alpha v\|_{L^\infty(G)} < \infty \right\},$$

$$W_\beta^{k,\infty}(G) := \left\{ v \in \mathcal{D}'(G) : \|v\|_{W_\beta^{k,\infty}(G)} := \max_{|\alpha|\leq k} \|r^\beta D^\alpha v\|_{L^\infty(G)} < \infty \right\}.$$

Definition 2.11. We define the function spaces

$$C(G) := \{v : G \to \mathbb{R} : v \text{ continuous}\}$$

and

$$C(\bar{G}) := \{v \in C(G) : v \text{ has a continuous extension to } \bar{G}\}.$$

The space $C(\bar{G})$ is a Banach space with the norm

$$\|v\|_{C(\bar{G})} := \sup_{x \in \bar{G}} |v(x)|.$$

Let $0 < \sigma \leq 1$. The space of *Hölder continuous* functions is defined as

$$C^{0,\sigma}(\bar{G}) := \{v \in C(\bar{G}) : v \ \sigma\text{-Hölder continuous for } |\alpha| = k\}.$$

The space $C^{0,\sigma}(\bar{G})$ is a Banach space with the norm

$$\|v\|_{C^{0,\sigma}(\bar{G})} := \|v\|_{C(\bar{G})} + \sup_{x,y \in \bar{G}, \ x \neq y} \frac{|v(x) - v(y)|}{|x - y|^\sigma}.$$

We set $C^{0,0}(\bar{G}) := C(\bar{G})$.

Definition 2.12. Let G be partitioned in disjoint, nonempty and open subdomains G_i, $i = 1, \ldots, n$. We define for $k \in \mathbb{N}_0$ and $p \in [1, \infty)$ the spaces

$$\mathcal{W}^{k,p}(G) := \prod_{i=1}^n W^{k,p}(G_i),$$

i.e., $v \in \mathcal{W}^{k,p}(G)$ if and only if $v_i := v|_{G_i} \in W^{k,p}(G_i)$ for all $i = 1, \ldots, n$. The corresponding norm is defined as

$$\|v\|_{\mathcal{W}^{k,p}(G)} := \left(\sum_{i=1}^n \|v_i\|_{W^{k,p}(G_i)}^p \right)^{1/p}.$$

Analogically, the corresponding seminorm is defined as

$$|v|_{\mathcal{W}^{k,p}(G)} := \left(\sum_{i=1}^{n} |v_i|_{W^{k,p}(G_i)}^{p} \right)^{1/p}.$$

Furthermore, we define

$$\mathcal{W}^{k,\infty}(G) := \{ v \in \mathcal{D}'(G) \, : \, v_i \in W^{k,\infty}(G_i) \}$$

with the norm and seminorm, respectively,

$$\|v\|_{\mathcal{W}^{k,\infty}(G)} := \max_{i=1,\ldots,n} \|v_i\|_{W^{k,\infty}(G_i)}, \quad |v|_{\mathcal{W}^{k,\infty}(G)} := \max_{i=1,\ldots,n} |v_i|_{W^{k,\infty}(G_i)}.$$

We introduce for $k \in \mathbb{N}_0$, $p \in [1,\infty)$ and $\beta \in \mathbb{R}$ the weighted spaces

$$\mathcal{V}_{\beta}^{k,p}(G) := \prod_{i=1}^{n} V_{\beta}^{k,p}(G_i),$$

with the norm and seminorm, respectively,

$$\|v\|_{\mathcal{V}_{\beta}^{k,p}(G)} := \left(\sum_{i=1}^{n} \|v_i\|_{V_{\beta}^{k,p}(G_i)}^{p} \right)^{1/p}, \quad |v|_{\mathcal{V}_{\beta}^{k,p}(G)} := \left(\sum_{i=1}^{n} |v_i|_{V_{\beta}^{k,p}(G_i)}^{p} \right)^{1/p}.$$

Moreover, we introduce

$$\mathcal{V}_{\beta}^{k,\infty}(G) := \{ v \in \mathcal{D}'(G) \, : \, v_i \in V_{\beta}^{k,\infty}(G_i) \}$$

and the norm and seminorm, respectively,

$$\|v\|_{\mathcal{V}_{\beta}^{k,\infty}(G)} := \max_{i=1,\ldots,n} \|v_i\|_{V_{\beta}^{k,\infty}(G_i)}, \quad |v|_{\mathcal{V}_{\beta}^{k,\infty}(G)} := \max_{i=1,\ldots,n} |v_i|_{V_{\beta}^{k,\infty}(G_i)}.$$

We define the spaces

$$\mathcal{C}^{0,\sigma}(\bar{G}) := \{ v : G \to \mathbb{R} \, : \, v_i \in C^{0,\sigma}(\bar{G}_i) \}$$

with the norm

$$\|v\|_{\mathcal{C}^{0,\sigma}(\bar{G})} := \max_{i=1,\ldots,n} \|v\|_{C^{0,\sigma}(\bar{G}_i)}.$$

We recall the Sobolev Embedding Theorem.

Theorem 2.13. *Let $k \in \mathbb{N}_0$ and $1 \le p < \infty$.*

(1) For $q \ge 1$, $l \in \mathbb{N}_0$ and $k \ge l$ with $k - d/p \ge l - d/q$ one has the continuous embedding

$$W^{k,p}(G) \hookrightarrow W^{l,q}(G).$$

The embedding is compact if $k > l$ and $k - d/p > l - d/q$.

(2) For $0 < \sigma < 1$ and $k - d/p \geq \sigma$ one has the continuous embedding

$$W^{k,p}(G) \hookrightarrow C^{0,\sigma}(\bar{G}).$$

The embedding is compact if $0 \leq \sigma \leq 1$ and $k - d/p > \sigma$.

The following trace theorem is proved for instance in [128, Theorem 8.7]

Theorem 2.14. *For $1/2 < s \leq 1$ there exists a linear and continuous trace operator*

$$T_0 : H^s(G) \to H^{s-1/2}(\partial G),$$

i.e., the inequality

$$\|T_0 v\|_{H^{s-1/2}(\partial G)} \leq c\|v\|_{H^s(G)} \quad \text{for } v \in H^s(G)$$

holds.

The following lemma is well known, see e.g. [14]. It concerns embedding results for the spaces $V_\beta^{2,2}(G)$ and $V_\beta^{2,\infty}(G)$, respectively.

Lemma 2.15. *The embeddings*

$$V_\beta^{2,2}(G) \hookrightarrow V_\gamma^{2,2}(G) \quad \text{for } \beta < \gamma, \tag{2.1}$$

$$V_\gamma^{2,\infty}(G) \hookrightarrow L^\infty(G) \quad \text{for } \gamma \leq 2 \tag{2.2}$$

hold.

Proof. Since $\beta < \gamma$ the embedding (2.1) follows directly from the definition of the spaces. For $u \in V_\gamma^{2,\infty}(G)$ one has $r^{\gamma-2}u \in L^\infty(G)$. From the fact that $\gamma \leq 2$ one obtains $u \in L^\infty(G)$ what proves (2.2). $\qquad\square$

The following two lemmas concerning the spaces $W_\beta^{k,p}(G)$ were originally published in [17]. The first one entails an embedding result whereas in the second one a norm equivalence is proved, that will be useful in the forthcoming derivation of a local interpolation error estimate (comp. Chapter 3).

Lemma 2.16. *For $p \in (1,\infty)$, $\beta > 1 - 2/p$ and $k \geq 0$ one has the compact embedding*

$$W_\beta^{k+1,p}(G) \overset{c}{\hookrightarrow} W_\beta^{k,p}(G). \tag{2.3}$$

For $p \in (1,\infty)$, $k \geq 1$ and $\beta \in (1 - 2/p, 1]$ the continuous embeddings

$$W_\beta^{k,p}(G) \hookrightarrow W_{\beta-1}^{k-1,p}(G) \hookrightarrow W^{k-1,p}(G) \hookrightarrow L^p(G) \hookrightarrow L^1(G) \tag{2.4}$$

are valid.

Proof. From Lemma 1.8 in [113] one has $W_\beta^{1,p}(G) \hookrightarrow V_\beta^{1,p}(G)$ for $\beta > 1 - 2/p$. Lemma 1.2 in [113] yields $V_\beta^{1,p}(G) \overset{c}{\hookrightarrow} W_{\beta-1}^{0,p}(G)$. Since $W_{\beta-1}^{0,p}(G) \hookrightarrow W_\beta^{0,p}(G)$ this shows the embedding $W_\beta^{1,p}(G) \overset{c}{\hookrightarrow} W_\beta^{0,p}(G)$. Applying this embedding to derivatives, one can conclude (2.3). The embedding $W_\beta^{k,p}(G) \hookrightarrow W_{\beta-1}^{k-1,p}(G)$ follows from Theorem 1.3 in [113]. The other embeddings in (2.4) can be concluded directly since $\beta \leq 1$ and $p > 1$. $\qquad\square$

Lemma 2.17. *For $p > 1$, $\beta \in (1 - 2/p, 1]$ and a function $v \in W_\beta^{k+1,p}(G)$ one has the norm equivalence*

$$\|v\|_{W_\beta^{k+1,p}(G)} \sim |v|_{W_\beta^{k+1,p}(G)} + \sum_{|\alpha| \leq k} \left| \int_G D^\alpha v \right|.$$

Proof. The following proof is based on an idea of [97, Chap. 4, §5]. Since one has for $p > 1$ and $\beta \in (1 - 2/p, 1]$ the embedding $W_\beta^{1,p}(G) \hookrightarrow L^1(G)$ (see (2.4)), the inequality

$$\|v\|_{W_\beta^{k+1,p}(G)} \geq c \left(|v|_{W_\beta^{k+1,p}(G)} + \sum_{|\alpha| \leq k} \left| \int_G D^\alpha v \right| \right)$$

holds. In order to show the other direction,

$$\|v\|_{W_\beta^{k+1,p}(G)} \leq c \left(|v|_{W_\beta^{k+1,p}(G)} + \sum_{|\alpha| \leq k} \left| \int_G D^\alpha v \right| \right), \tag{2.5}$$

we use a proof by contradiction. If inequality (2.5) was not valid, then there would be a sequence (v_n) with $v_n \in W_\beta^{k+1,p}(G)$ such that

$$\|v_n\|_{W_\beta^{k+1,p}(G)} = 1, \tag{2.6}$$

$$|v_n|_{W_\beta^{k+1,p}(G)} + \sum_{|\alpha| \leq k} \left| \int_G D^\alpha v_n \right| \leq \frac{1}{n}. \tag{2.7}$$

Since (v_n) is a bounded sequence in $W_\beta^{k+1,p}(G)$ and $W_\beta^{k+1,p}(G) \overset{c}{\hookrightarrow} W_\beta^{k,p}(G)$ (see (2.3)) there is a convergent subsequence $(v_{n_l}) \in W_\beta^{k,p}(G)$. In the following we suppress the index l and write (v_n) for this subsequence. Because of the completeness of $W_\beta^{k,p}(G)$ there is a function $v \in W_\beta^{k,p}(G)$, such that

$$\|v - v_n\|_{W_\beta^{k,p}(G)} \overset{n\to\infty}{\longrightarrow} 0. \tag{2.8}$$

With (2.7) one can conclude $|v_n|_{W_\beta^{k+1,p}(G)} \leq 1/n$, what results in

$$|v_n|_{W_\beta^{k+1,p}(G)} \overset{n\to\infty}{\longrightarrow} 0. \tag{2.9}$$

In the following we show that (v_n) is a Cauchy sequence in $W_\beta^{k+1,p}(G)$. For a fixed and arbitrary small $\varepsilon > 0$ and numbers n, m large enough one obtains with (2.8) and (2.9)

$$\|v_n - v_m\|_{W_\beta^{k+1,p}(G)}^p = \|v_n - v_m\|_{W_\beta^{k,p}(G)}^p + |v_n - v_m|_{W_\beta^{k+1,p}(G)}^p$$

$$\leq \frac{\varepsilon}{3} + C|v_n|_{W_\beta^{k+1,p}(G)}^p + C|v_m|_{W_\beta^{k+1,p}(G)}^p$$

$$\leq \frac{\varepsilon}{3} + \frac{\varepsilon}{3} + \frac{\varepsilon}{3} = \varepsilon.$$

Since $W_\beta^{k+1,p}(G)$ is complete, there is a function $v^* \in W_\beta^{k+1,p}(G)$ with

$$\|v_n - v^*\|_{W_\beta^{k+1,p}(G)} \xrightarrow{n \to \infty} 0$$

and one arrives with (2.6) at

$$\|v^*\|_{W_\beta^{k+1,p}(G)} = 1. \tag{2.10}$$

Furthermore one can conclude from (2.7)

$$|v^*|_{W_\beta^{k+1,p}(G)} + \sum_{|\alpha| \leq k} \left| \int_G D^\alpha v^* \right| = 0, \tag{2.11}$$

in particular $|v^*|_{W_\beta^{k+1,p}(G)} = 0$, which means that $D^\alpha v^* = 0 \quad \forall \alpha : |\alpha| = k+1$, that is v^* is a polynomial over G with degree at most k. Since v^* also fulfills (2.11), it follows directly $v^* = 0$. This is a contradiction to (2.10), what proves (2.5). $\square$

2.3 Graded triangulations

In this section we introduce triangulations where the element sizes depend on the distance of the element to a corner or an edge. All triangulations $\mathcal{T}_h = \{T\}$ of Ω, that we consider, are admissible in Ciarlet's sense [42], i.e.,

- $\bar{\Omega} = \bigcup_{T \in \mathcal{T}_h} \bar{T}$
- For two arbitrary elements $T_1, T_2 \in \mathcal{T}_h$ with $T_1 \neq T_2$ one has $T_1 \cap T_2 = \emptyset$.
- Any face of any element $T_1 \in \mathcal{T}_h$ is either a subset of the boundary $\partial\Omega$ or a face of another element $T_2 \in \mathcal{T}_h$.

2.3.1 Two-dimensional domain

Let us first consider a two-dimensional, bounded and polygonal domain Ω. We assume that Ω has only one corner with interior angle $\omega > \omega_0$ located at the origin. The critical angle ω_0 depends on the problem under consideration, e.g. for the Poisson problem and

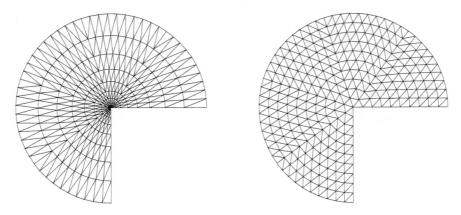

Figure 2.1: Graded mesh with $\mu = 0.4$ (left) and quasi-uniform mesh ($\mu = 1.0$)

L^2-error estimates it is $\omega_0 = \pi$, for the corresponding L^∞-error estimates one has $\omega_0 = \pi/2$. For details on this, we refer to Chapter 4. With a global mesh parameter h, a grading parameter $\mu \in (0, 1]$ and the distance r_T of a triangle T to the corner,

$$r_T := \inf_{(x_1, x_2) \in T} \sqrt{x_1^2 + x_2^2},$$

we assume that the element size $h_T := \mathrm{diam}T$ satisfies

$$
\begin{aligned}
c_1 h^{1/\mu} \leq h_T \leq c_2 h^{1/\mu} \qquad &\text{for } r_T = 0 \\
c_1 h r_T^{1-\mu} \leq h_T \leq c_2 h r_T^{1-\mu} \qquad &\text{for } r_T > 0.
\end{aligned}
\tag{2.12}
$$

Notice, that the number of elements of such a triangulation is of order h^{-2}, see e.g. [16]. In Figure 2.1 one can see such isotropic graded meshes for different values of μ.

2.3.2 Three-dimensional domain

We consider a prismatic domain $\Omega = G \times Z$, where $G \subset \mathbb{R}^2$ is a bounded polygonal domain and $Z := (0, z_0) \subset \mathbb{R}$ is an interval. It is assumed that the cross-section G has only one corner with interior angle $\omega > \pi$ at the origin. For the construction of a mesh in Ω, we first introduce a graded triangulation $\{\tau\}$ in the two-dimensional domain G according to (2.12). This means the element size $h_\tau := \mathrm{diam}\tau$ satisfies

$$
\begin{aligned}
c_1 h^{1/\mu} \leq h_\tau \leq c_2 h^{1/\mu} \qquad &\text{for } r_\tau = 0 \\
c_1 h r_\tau^{1-\mu} \leq h_\tau \leq c_2 h r_\tau^{1-\mu} \qquad &\text{for } r_\tau > 0.
\end{aligned}
$$

where r_τ is the distance to the corner,

$$r_\tau := \inf_{(x_1, x_2) \in \tau} \sqrt{x_1^2 + x_2^2},$$

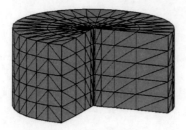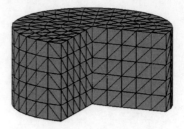

Figure 2.2: Anisotropic graded mesh with $\mu = 0.4$ (left) and uniform mesh ($\mu = 1.0$)

μ the grading parameter and h the global mesh parameter. From this graded two-dimensional mesh a three-dimensional mesh of pentahedra is built by extruding the triangles τ in x_3-direction with uniform mesh size h. In order to generate an anisotropic graded tetrahedral mesh, each of these pentahedra is divided into tetrahedra. We can characterize the elements T of such a mesh by the three mesh sizes $h_{T,1}$, $h_{T,2}$ and $h_{T,3}$, where $h_{T,i}$ is the length of the projection of T on the x_i-axis, $i = 1, 2, 3$. In detail, with r_T being the distance of the element T to the edge,

$$r_T := \inf_{x \in T} \sqrt{x_1^2 + x_2^2},$$

the element sizes satisfy

$$
\begin{aligned}
c_1 h^{1/\mu} \leq h_{T,i} \leq c_2 h^{1/\mu} &\qquad \text{for } r_T = 0, \\
c_1 h r_T^{1-\mu} \leq h_{T,i} \leq c_2 h r_T^{1-\mu} &\qquad \text{for } r_T > 0, \\
c_1 h \leq h_{T,3} \leq c_2 h, &
\end{aligned}
\tag{2.13}
$$

for $i = 1, 2$. The number of elements is of order h^{-3} and therefore asymptotically not increasing in comparison with a quasi-uniform mesh. Figure 2.2 shows such anisotropic graded tetrahedral meshes for $\mu = 0.4$ and $\mu = 1.0$.

CHAPTER **3**

Interpolation of nonsmooth functions

3.1 Suitable interpolation operators

In this chapter we consider estimates of the approximation error for a quasi-interpolant on anisotropic finite element meshes in prismatic domains. Such estimates are main ingredients of the error analysis for a finite element discretization.

Let us discuss shortly why the use of quasi-interpolation operators gains some advantage for anisotropic, three-dimensional finite elements. If one takes the standard Lagrangian (nodal) interpolation operator, which uses nodal values of the function for the definition of the interpolant, one can benefit from the very useful property, that Dirichlet boundary conditions are preserved by this operator. However, it was shown in [7] for an anisotropic, three-dimensional finite element T, that the local interpolation error estimate

$$|u - I_h u|_{W^{1,p}(T)} \leq c \sum_{i=1}^{3} h_{i,T} \left| \frac{\partial u}{\partial x_i} \right|_{W^{1,p}(T)} \tag{3.1}$$

and even its simplified version $|u - I_h u|_{W^{1,p}(T)} \leq c \max_i h_{i,T} \, |u|_{W^{2,p}(T)}$ is valid under the condition $p > 2$ only. The main drawback of this result is, that it allows an estimate of the finite element error in $H^1(\Omega)$ only for a right-hand side in $L^p(\Omega)$, $p > 2$. The consequence is, that one cannot achieve an $L^2(\Omega)$-error estimate via the Aubin-Nitsche method. A way out is the use of quasi-interpolation operators as for example introduced in [44, 120]. The basic idea is to replace nodal values by suitable averaged values. Apel investiged in [6] several quasi-interpolation operators for anisotropic elements. It turned out that the classical operators introduced in [44, 120] are not uniformly $W^{1,p}$-stable in the aspect ratio and do not satisfy an estimate like (3.1) (with T replaced by a patch S_T on the right hand side). As a remedy he introduced three modifications of the Scott-Zhang interpolant for which such estimates hold. As the original ones these modified operators have the disadvantage that they preserve Dirichlet boundary conditions on parts of the boundary only. In the following we introduce another modification tailored to pure Dirichlet problems. This interpolant is closely related to the operator E_h introduced by Apel in [6]. Although some of our proofs are applications or simple extensions of derivations in that paper, the estimate of an occurring additional error term requires new ideas for the proof. Furthermore we prove local estimates for the operator E_h in the space $W_\beta^{k,p}(\Omega)$ such that the results can also be used for the derivation of global error estimates for the Neumann problem. The following results were originally published in [17].

3.2 Tensor product meshes

Let Ω be a prismatic domain, i.e. $\Omega = G \times Z \subset \mathbb{R}^3$, where $G \subset \mathbb{R}^2$ is a bounded polygonal domain and $Z := (0, z_0) \subset \mathbb{R}$ is an interval. The different parts of the boundary are denoted by

$$\Gamma_B := \{x \in \partial\Omega \ : \ x_3 = 0 \text{ or } x_3 = z_0\} \quad \text{and} \quad \Gamma_M := \partial\Omega \backslash \Gamma_B.$$

The crosssection G is assumed to have only one corner with interior angle $\omega > \pi$ at the origin; thus Ω has only one "singular edge" which is part of the x_3-axis. For a triangulation of Ω we do not demand the elements to be shape-regular. In contrast we are interested in *anisotropic elements*. According to [9], we consider the four reference elements

$$\hat{T}_1 := \{(\hat{x}_1, \hat{x}_2, \hat{x}_3) \in \mathbb{R}^3 : 0 < \hat{x}_1 < 1, 0 < \hat{x}_2 < 1 - \hat{x}_1, 0 < \hat{x}_3 < 1 - \hat{x}_1 - \hat{x}_2\},$$
$$\hat{T}_2 := \{(\hat{x}_1, \hat{x}_2, \hat{x}_3) \in \mathbb{R}^3 : 0 < \hat{x}_1 < 1, 0 < \hat{x}_2 < 1 - \hat{x}_1, \hat{x}_1 < \hat{x}_3 < 1\},$$
$$\hat{T}_3 := \{(\hat{x}_1, \hat{x}_2, \hat{x}_3) \in \mathbb{R}^3 : 0 < \hat{x}_1 < 1, 0 < \hat{x}_2 < \hat{x}_1, 0 < \hat{x}_3 < \hat{x}_1 - \hat{x}_2\},$$
$$\hat{T}_4 := \{(\hat{x}_1, \hat{x}_2, \hat{x}_3) \in \mathbb{R}^3 : 0 < \hat{x}_1 < 1, 0 < \hat{x}_2 < \hat{x}_1, 1 - \hat{x}_1 < \hat{x}_3 < 1\}.$$

For an illustration we refer to Figure 3.1. For elements with a face parallel to the $x_1 - x_2$-plane we use $\hat{T}_1$ and $\hat{T}_3$, for elements without such a face $\hat{T}_2$ and $\hat{T}_4$ are considered. Elements with exactly one vertex with $r = 0$ are mapped to $\hat{T}_3$ or $\hat{T}_4$, in all other cases (zero or two vertices with $r = 0$) $\hat{T}_1$ and $\hat{T}_2$ are used In the following we refer to the suitable reference element by $\hat{T}$. In order to be able to write down our proofs in a concise way, we restrict ourselves first to *tensor product meshes*. According to [6] an affine finite element is called *tensor product element*, when the transformation of a reference element $\hat{T}$ to the element T has the form

$$\begin{pmatrix} x_1 \\ x_2 \\ x_3 \end{pmatrix} = \begin{pmatrix} h_{1,T} & 0 & 0 \\ 0 & h_{2,T} & 0 \\ 0 & 0 & h_{3,T} \end{pmatrix} \begin{pmatrix} \hat{x}_1 \\ \hat{x}_2 \\ \hat{x}_3 \end{pmatrix} + b_T,$$

where $b_T \in \mathbb{R}^3$. Note that the vertices of a tensor element are located in the corners of a cuboid with edge lengths $h_{1,T}$, $h_{2,T}$ and $h_{3,T}$. We explain in Subsection 3.6, how the results extend to a more general mesh type.

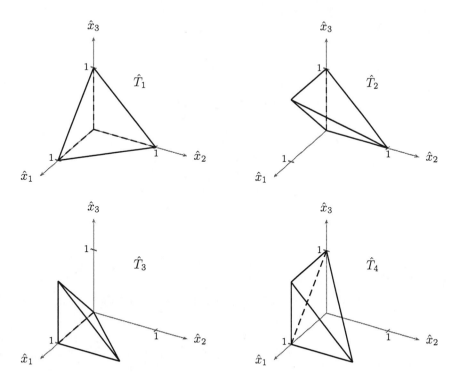

Figure 3.1: Reference elements for anisotropic interpolation error estimates

In addition we demand that there is no rapid change in the element sizes, this means, that the relation

$$h_{i,T} \sim h_{i,T'} \text{ for all } T' \text{ with } \overline{T} \cap \overline{T'} \neq \emptyset$$

holds for $i = 1, 2, 3$. Furthermore we define the set

$$M_T := \text{int} \bigcup_{i \in I_T} \overline{T_i},$$

where the set I_T contains all indices i for which $\overline{T_i} \cap \overline{T} \neq \emptyset$ and the projection of T_i on the x_1x_2-plane is the same as the one of T. With S_T we denote the smallest triangular prism that contains M_T. Notice that the height of S_T is in the order of $h_{3,T}$. We further define

$$S_{\hat{T}} := \left\{ (\hat{x}_1, \hat{x}_2, \hat{x}_3) \in \mathbb{R}^3 : 0 < \hat{x}_1 < 1, 0 < \hat{x}_2 < 1 - \hat{x}_1, 0 < \hat{x}_3 < 1 \right\}$$

as reference patch.

21

3.3 Quasi-Interpolation operators

We define the Scott-Zhang type interpolant $E_h : W^{l,p}(\Omega) \to V_h$ by

$$(E_h u)(x) := \sum_{i \in I} a_i \varphi_i(x), \tag{3.2}$$

which was originally introduced in [6]. It is $l \geq 2$ for $p = 1$ and $l > 2/p$ otherwise, compare also the forthcoming Remark 3.1. The set I is the index set of all nodes, the functions φ_i, $i \in I$, are *nodal basis functions*, i.e. $\varphi_i(X_j) = \delta_{ij}$ for all $i, j \in I$, where $X_i = (X_{i,1}, X_{i,2}, X_{i,3}) \in \mathbb{R}^3$ are the nodes of the finite element mesh. In order to specify a_i, we first introduce the subsets $\sigma_i \subset \bar{\Omega}$ by the following properties.

(P1) σ_i is one-dimensional and parallel to the x_3-axis.

(P2) $X_i \in \bar{\sigma}_i$

(P3) There exists an edge e of some element T such that the projection of e on the x_3-axis coincides with the projection of σ_i.

(P4) If the projections of any two points X_i and X_j on the x_3-axis coincide then so do the projections of σ_i and σ_j.

Note that the properties (P3) and (P4) make sense since we consider tensor product meshes. Now a_i is chosen as the value of the $L^2(\sigma_i)$-projection of u in the space of linear functions over $\sigma_i \subset \bar{\Omega}$ at the node X_i,

$$a_i := (Q_{\sigma_i} u)(X_i)$$

with

$$Q_{\sigma_i} : L^2(\sigma_i) \to \mathcal{P}_{1,\sigma_i}$$

where $\mathcal{P}_{1,\sigma_i}$ is the space of polynomials over σ_i with a degree of at most 1.

We denote by $\Phi_{0,i}$ and $\Phi_{1,i}$ the two one-dimensional linear nodal functions corresponding to $\sigma_i = \overrightarrow{X_i X_j}$, that means

$$\Phi_{0,i}(X_{i,3}) = 1, \qquad \Phi_{0,i}(X_{j,3}) = 0,$$
$$\Phi_{1,i}(X_{i,3}) = 0, \qquad \Phi_{1,i}(X_{j,3}) = 1.$$

Besides we define $\Psi_{0,i}$ and $\Psi_{1,i}$ as the two linear functions, that are biorthogonal to $\{\Phi_{0,i}, \Phi_{1,i}\}$,

$$\int_{\sigma_i} \Phi_{k,i} \Psi_{l,i} = \delta_{k,l} \quad (k, l = 0, 1). \tag{3.3}$$

Notice that $\Phi_{k,i}$ depends only on $X_{i,3}$, what means that $\Phi_{k,i} = \Phi_{k,m}$ if $X_{i,3} = X_{m,3}$, $k = 0, 1$. The same is valid for $\Psi_{k,i}$. With this setting we can write the interpolation operator E_h as

$$E_h u(x) = \sum_{i \in I} (Q_{\sigma_i} u)(X_i) \varphi_i(x)$$

$$= \sum_{i \in I} \left[\Phi_{0,i}(X_{i,3}) \int_{\sigma_i} u \Psi_{0,i} \, ds + \Phi_{1,i}(X_{i,3}) \int_{\sigma_i} u \Psi_{1,i} \, ds \right] \varphi_i(x)$$

$$= \sum_{i \in I} \left[\int_{\sigma_i} u \Psi_{0,i} \, ds \right] \varphi_i(x). \tag{3.4}$$

Remark 3.1. $E_h u$ is well-defined only for $u \in W^{l,p}(\Omega)$ with

$$l \geq 2 \quad \text{for } p = 1, \qquad l > \frac{2}{p} \quad \text{otherwise.}$$

This guarantees $u|_{\sigma_i} \in L^1(\Omega)$. In the special case that $u \in W^{2,2}_{1-\pi/\omega+\varepsilon}(\Omega)$ the interpolant $E_h u$ is also well-defined since one has the embedding $W^{2,2}_{1-\pi/\omega+\varepsilon}(\Omega) \hookrightarrow W^{1+\pi/\omega-\varepsilon,2}_0(\Omega)$ (see [113], Theorem 1.3) and $1 + \pi/\omega - \varepsilon > 1$.

The disadvantage of E_h is, that it preserves Dirichlet boundary conditions on Γ_M only, but not on Γ_B. But this is necessary in order to derive an estimate for the finite element error for pure Dirichlet problems. In order to be able to treat boundary value problems with Dirichlet boundary conditions on the whole boundary $\partial\Omega$, we introduce the operator E_{0h} as modification of E_h.

Let J be the index set, which includes the indices of all nodes not belonging to Γ_B and

$$V_h := \{ v_h \in H^1(\Omega) : v_h|_T \in \mathcal{P}_1 \text{ for all } T \in \mathcal{T}_h \},$$
$$V_{0h} := \{ v_h \in V_h : v_h|_{\partial\Omega} = 0 \}.$$

We define $E_{0h} : W^{2,p}(\Omega) \to V_{0h}$ as

$$(E_{0h} u)(x) := \sum_{i \in J} (Q_{\sigma_i} u)(X_i) \varphi_i(x). \tag{3.5}$$

Since $\varphi_i(x) = 0$ for all $x \in \Gamma_B$ and $i \in J$, the operator E_{0h} is preserving homogeneous Dirichlet boundary conditions also on Γ_B.

In the following we assume

$$h_{1,T} \leq h_{2,T} \leq h_{3,T} \tag{3.6}$$

without loss of generality.

3.4 Local estimates in classical Sobolev spaces

We first recall an approximation result from [6].

Theorem 3.2. *Consider an element T of a tensor product mesh and assume that (3.6) is fulfilled. Then the approximation error estimate*

$$|u - E_h u|_{W^{1,q}(T)} \leq c|T|^{1/q-1/p} \sum_{|\alpha|=1} h_T^\alpha |D^\alpha u|_{W^{1,p}(S_T)} \tag{3.7}$$

holds for $p \in [1,\infty]$, q such that $W^{2,p}(T) \hookrightarrow W^{1,q}(T)$ and $u \in W^{2,p}(S_T)$.

Proof. If one sets $l = 2$, $m = 1$ formula (3.7) is exactly formula (6.6) in Theorem 10 of [6]. □

Our aim is now to estimate $|u - E_{0h}u|_{W^{1,q}(T)}$ for a function $u \in W^{2,p}(T)$, $p \in [1,\infty]$, q such that $W^{2,p}(T) \hookrightarrow W^{1,q}(T)$ and $u|_{\Gamma_B} = 0$. With the triangle inequality we get

$$|u - E_{0h}u|_{W^{1,q}(T)} \leq |u - E_h u|_{W^{1,q}(T)} + |E_h u - E_{0h}u|_{W^{1,q}(T)}. \tag{3.8}$$

The first term on the right-hand side is treated in Theorem 3.2. It remains to find an estimate for the second term. To this end, we first prove the following auxiliary result.

Lemma 3.3. *Let T be an element with $\overline{T} \cap \Gamma_B \neq \emptyset$, I the index set of the nodes in $\overline{T} \cap \Gamma_B$ and u a function in $W^{2,p}(S_T)$ with S_T as defined in Section 3.2, $p \in [1,\infty]$ and with $u|_{\Gamma_B} = 0$. Then for every $i \in I$ and every linear function $\tilde{\Phi}_{1,i}$ with $\tilde{\Phi}_{1,i}|_{\sigma_i} = \Phi_{1,i}$ and $\tilde{\Phi}_{1,i}|_{\Gamma_B} = 0$ there exists $c_i \in \mathbb{R}$, such that*

$$\sum_{|\alpha|\leq 2} h^\alpha \left\| D^\alpha \left(u - c_i \tilde{\Phi}_{1,i} \right) \right\|_{L^p(S_T)} \leq c \sum_{|\alpha|=2} h^\alpha \|D^\alpha u\|_{L^p(S_T)}. \tag{3.9}$$

Furthermore one has

$$\sum_{|\alpha|\leq 1} h^\alpha |D^\alpha u|_{W^{1,p}(S_T)} \leq c \sum_{|\alpha|=1} h^\alpha |D^\alpha u|_{W^{1,p}(S_T)}. \tag{3.10}$$

Proof. Let g be a continuous function with the properties of a norm, i.e.

$$g(t_1,\ldots,t_n) \geq 0 \text{ and } g((t_1,\ldots,t_n)) = 0 \Leftrightarrow t_1 = \cdots = t_n = 0,$$
$$g(\lambda t_1,\ldots,\lambda t_n) = |\lambda| g((t_1,\ldots,t_n)),$$
$$g(t_1 + \tau_1,\ldots,t_n + \tau_n) \leq g(t_1,\ldots,t_n) + g(\tau_1,\ldots,\tau_n).$$

In Theorem 4.5.1 of [97] it is shown that for such functions and for linear functionals $l_1, l_2, \ldots, l_N$ that are bounded in $W^{k,p}(\Omega)$ and do not vanish simultaneously on a polynomial with degree less than k besides the zero polynomial, the inequality

$$\|u\|_{W^{k,p}(\Omega)} \leq c \left(g(l_1 u, l_2 u, \ldots, l_N u) + |u|_{W^{k,p}(\Omega)} \right) \tag{3.11}$$

is valid. Here N is the number of independent monomials of degree $\leq k - 1$.

Now we prove (3.9) and (3.10) for the reference patch $S_{\hat{T}}$. In our case there is $N = 4$, what is the number of monomials of degree less than or equal to 1 in three dimensions. We denote by $\hat{e}_i$ ($i = 1, 2, 3$) the three edges of $S_{\hat{T}}$ in the x_1x_2-plane. Then we set $l_iv := \int_{\hat{e}_i} v$, $i = 1, 2, 3$ and $l_4v := \int_{S_{\hat{T}}} v$. For g we choose $g(t_1, t_2, t_3, t_4) = \sum_{i=1}^{4} |t_i|$. Now we set c_i such that $\int_{S_{\hat{T}}} \left(\hat{u} - c_i \hat{\Phi}_{1,i} \right) = 0$ and we get

$$\|\hat{u} - c_i \hat{\Phi}_{1,i}\|_{W^{2,p}(S_{\hat{T}})} \leq c \left(\sum_{j=1}^{3} \left| \int_{\hat{e}_j} (\hat{u} - c_i \hat{\Phi}_{1,i}) \right| \right.$$
$$\left. + \left| \int_{S_{\hat{T}}} \left(\hat{u} - c_i \hat{\Phi}_{1,i} \right) \right| + |\hat{u} - c_i \hat{\Phi}_{1,i}|_{W^{2,p}(S_{\hat{T}})} \right).$$

Since $\hat{\Phi}_{1,i}$ is linear and $\hat{\Phi}_{1,i}|_{\hat{e}_j} = 0$ ($j = 1, 2, 3$) we end up with

$$\|\hat{u} - c_i \hat{\Phi}_{1,i}\|_{W^{2,p}(S_{\hat{T}})} \leq c\, |\hat{u}|_{W^{2,p}(S_{\hat{T}})}.$$

The transformation back to S_T yields assertion (3.9).

In the case of (3.10) we have $k = 1$ and $N = 1$. We set $l_1v = \int_{S_{\hat{T}} \cap \{z=0\}} v\, ds$. Since $\hat{u}$ vanishes on $S_{\hat{T}} \cap \{z = 0\}$ one has $l_1\hat{u} = 0$ and with (3.11) this yields

$$\|\hat{u}\|_{W^{1,p}(S_{\hat{T}})} \leq c\, |\hat{u}|_{W^{1,p}(S_{\hat{T}})}.$$

The transformation back to S_T results in inequality (3.10). $\qquad\square$

With this result at hand, we are now able to give an estimate of the second term of the right-hand side of inequality (3.8).

Theorem 3.4. *Consider an element T of a tensor product mesh and assume that condition (3.6) is fulfilled. Then the error estimate*

$$|E_{0h}u - E_hu|_{W^{1,q}(T)} \leq c\, |T|^{1/q-1/p} \sum_{|\alpha|=1} h^\alpha |D^\alpha u|_{W^{1,p}(S_T)} \tag{3.12}$$

holds if $p \in [1, \infty]$, q is such that $W^{2,p}(T) \hookrightarrow W^{1,q}(T)$, $u \in W^{2,p}(S_T)$ and $u|_{\overline{T} \cap \Gamma_B} = 0$.

Proof. For an element T with $\overline{T} \cap \Gamma_B = \emptyset$ one has $E_{0h}u - E_hu = 0$ and (3.12) is valid. For an element T with $\overline{T} \cap \Gamma_B \neq \emptyset$ denote by B_T the index set of nodes belonging to Γ_B, $B_T := \{i : X_i \in \overline{T} \cap \Gamma_B\}$. We treat the derivatives in the different directions separately.

25

For the estimate of the derivative in x_3-direction it follows together with (3.4) and (3.3)

$$
\begin{aligned}
\left\| \frac{\partial}{\partial x_3}(E_h - E_{0h})u \right\|_{L^q(T)} &= \left\| \sum_{i \in B_T} (Q_{\sigma_i} u) \frac{\partial}{\partial x_3} \varphi_i \right\|_{L^q(T)} \\
&= \left\| \sum_{i \in B_T} \left[\int_{\sigma_i} u \Psi_{0,i} \right] \frac{\partial}{\partial x_3} \varphi_i \right\|_{L^q(T)} \\
&= \left\| \sum_{i \in B_T} \left[\int_{\sigma_i} (u - c_i \Phi_{1,i}) \Psi_{0,i} \right] \frac{\partial}{\partial x_3} \varphi_i \right\|_{L^q(T)}
\end{aligned}
\tag{3.13}
$$

for arbitrary $c_i \in \mathbb{R}$. We use

$$
\|\Psi_{0,i}\|_{L^\infty(\sigma_i)} \le c |\sigma_i|^{-1}
$$

and the trace theorem $W^{2,p}(S_T) \hookrightarrow L^1(\sigma_i)$, $p \ge 1$ in the form

$$
\|v\|_{L^1(\sigma_i)} \le c |\sigma_i| |T|^{-1/p} \sum_{|\alpha| \le 2} h^\alpha \|D^\alpha v\|_{L^p(S_T)}
$$

to get the estimate

$$
\begin{aligned}
\left| \int_{\sigma_i} (u - c_i \Phi_{1,i}) \Psi_{0,i} \, ds \right| &\le \|\Psi_{0,i}\|_{L^\infty(\sigma_i)} \|u - c_i \Phi_{1,i}\|_{L^1(\sigma_i)} \\
&\le c |T|^{-1/p} \sum_{|\alpha| \le 2} h^\alpha \|D^\alpha(u - c_i \tilde{\Phi}_{i,1})\|_{L^p(S_T)},
\end{aligned}
$$

where $\tilde{\Phi}_{i,1}$ is a linear function with $\tilde{\Phi}_{i,1}|_{\sigma_i} = \Phi_{i,1}$. With Lemma 3.3 we can conclude

$$
\left| \int_{\sigma_i} (u - c_i \Phi_{1,i}) \Psi_{0,i} \, ds \right| \le c |T|^{-1/p} \sum_{|\alpha| = 2} h^\alpha \|D^\alpha u\|_{L^p(S_T)}.
$$

Taking into account that

$$
\left\| \frac{\partial}{\partial x_3} \varphi_i \right\|_{L^q(T)} \le c |T|^{1/q} h_3^{-1} \quad \text{for } i \in B
$$

we can continue from equation (3.13) with

$$
\begin{aligned}
\left\| \frac{\partial}{\partial x_3}(E_h - E_{0h})u \right\|_{L^q(T)} &\le c \sum_{i \in B_T} \left(\left| \int_{\sigma_i} (u - c_i \Phi_{1,i}) \Psi_{0,i} \right| \left\| \frac{\partial}{\partial x_3} \varphi_i \right\|_{L^q(T)} \right) \\
&\le c |T|^{1/q - 1/p} h_3^{-1} \sum_{|\alpha| = 2} h^\alpha \|D^\alpha u\|_{L^p(S_T)}.
\end{aligned}
$$

With condition (3.6) we finally conclude

$$
\left\| \frac{\partial}{\partial x_3}(E_h - E_{0h})u \right\|_{L^q(T)} \le c |T|^{1/q - 1/p} \sum_{|\alpha| = 1} h^\alpha \|D^\alpha u\|_{W^{1,p}(S_T)}.
\tag{3.14}
$$

For the estimates concerning the derivatives in x_2- and x_1-direction we use a technique developed in [6]. Let us discuss the case of the x_2-derivative; the x_1-derivative can be proved by analogy.

First we consider the case that three nodes of T are contained in Γ_B, that means $|B_T| = 3$. We denote these nodes with X_0, X_1 and X_2, where the edge spanned by X_0 and X_1 is parallel to the x_1-axis and the one spanned by X_0 and X_2 parallel to the x_2-axis. Then one has

$$(E_{0h}u - E_h u)|_T = \sum_{i=0}^{2} a_i \varphi_i = (a_0 - a_2)\varphi_0 + a_2(\varphi_0 + \varphi_2) + a_1 \varphi_1$$

where we have set $a_i := \int_{\sigma_i} u \Psi_{0,i}$.

Taking into account that T is a tensor product element, we can conclude

$$\frac{\partial}{\partial x_2}\varphi_1 = 0 \quad \text{and} \quad \frac{\partial}{\partial x_2}(\varphi_0 + \varphi_2) = 0.$$

This yields

$$\left\| \frac{\partial}{\partial x_2}(E_{0h} - E_h)u \right\|_{L^q(T)} = |a_0 - a_2| \left\| \frac{\partial}{\partial x_2}\varphi_0 \right\|_{L^q(T)}. \tag{3.15}$$

Since $\{x_3 : (x_1, x_2, x_3) \in \sigma_0\} = \{x_3 : (x_1, x_2, x_3) \in \sigma_2\}$, $\Psi_{0,0} = \Psi_{0,2}$ and $X_{0,1} = X_{2,1}$ we get for the first factor

$$|a_0 - a_2| = \left| \int_{\sigma_0} u(X_{0,1}, X_{0,2}, z)\Psi_{0,0}(z)\, dz - \int_{\sigma_2} u(X_{0,1}, X_{2,2}, z)\Psi_{0,2}(z)\, dz \right|$$

$$= \left| \int_{\sigma_0} \Psi_{0,0}(z) \int_{X_{0,2}}^{X_{2,2}} \frac{\partial}{\partial x_2} u(X_{0,1}, y, z)\, dy\, dz \right|$$

$$\leq c \|\Psi_{0,0}\|_{L^\infty(\sigma_0)} \left| \int_{\sigma_0} \int_{X_{0,2}}^{X_{2,2}} \frac{\partial}{\partial x_2} u(X_{0,1}, y, z)\, dy\, dz \right|$$

$$\leq c h_1^{-1} h_3^{-1} \sum_{|\alpha| \leq 1} h^\alpha \left\| D^\alpha \left(\frac{\partial u}{\partial x_2} \right) \right\|_{L^1(S_T)}.$$

In the last estimate we have used the trace theorem $W^{1,1}(S_T) \hookrightarrow L^1(\Xi_1)$ where Ξ_1 is the two-dimensional manifold spanned by σ_0 and $X_0 X_2$ in the form

$$\|u\|_{L^1(\Xi_1)} \leq |\Xi_1||T|^{-1} \sum_{|\alpha| \leq 1} h^\alpha \|D^\alpha u\|_{L^1(S_T)}.$$

With

$$\left\| \frac{\partial}{\partial x_2}\varphi_0 \right\|_{L^q(T)} \leq c h_2^{-1} |T|^{1/q},$$

obtained by using the inverse inequality, it follows from (3.15) with the Hölder inequality

$$\left\| \frac{\partial}{\partial x_2}(E_{0h} - E_h)u \right\|_{L^q(T)} \leq c\,(h_1 h_2 h_3)^{-1}|T|^{1/q} \sum_{|\alpha|\leq 1} h^\alpha \left\| D^\alpha \left(\frac{\partial u}{\partial x_2} \right) \right\|_{L^1(S_T)}$$

$$\leq c\,|T|^{1/q-1/p} \sum_{|\alpha|\leq 1} h^\alpha \left\| D^\alpha \left(\frac{\partial u}{\partial x_2} \right) \right\|_{L^p(S_T)}.$$

The application of Lemma 3.3 yields

$$\left\| \frac{\partial}{\partial x_2}(E_{0h} - E_h)u \right\|_{L^q(T)} \leq c\,|T|^{1/q-1/p} \sum_{|\alpha|=1} h^\alpha \left\| D^\alpha \left(\frac{\partial u}{\partial x_2} \right) \right\|_{L^p(S_T)} \tag{3.16}$$

since $\frac{\partial u}{\partial x_2} = 0$ on Γ_B. Let us now consider the case where only two nodes X_0, X_1 of T are contained in Γ_B, what means $|B_T| = 2$. One has

$$(E_{0h}u - E_h u)|_T = a_0 \varphi_0 - a_1 \varphi_1. \tag{3.17}$$

We have to treat three different cases. First the case that the edge spanned by X_0 and X_1 is parallel to the x_2-axis, then the case that it is parallel to the x_1-axis and finally the case that is nor parallel to the x_1-axis nor to the x_2-axis. We first consider the case that the edge is parallel to the x_2-axis. One can rewrite (3.17) by

$$(E_{0h} - E_h u)|_T = (a_0 - a_1)\varphi_0 + a_1(\varphi_0 + \varphi_1).$$

Now one can proceed exactly as in the case with three nodes in Γ_B and obtain (3.16).

If the edge spanned by X_0 and X_1 is parallel to the x_1-axis one has

$$\frac{\partial}{\partial x_2}\varphi_0 = \frac{\partial}{\partial x_2}\varphi_1 = 0 \tag{3.18}$$

and from (3.17) one can conclude

$$\left\| \frac{\partial}{\partial x_2}(E_{0h} - E_h)u \right\|_{L^q(T)} = 0. \tag{3.19}$$

Consider now the case, where the edge spanned by X_0 and X_1 is neither parallel to the x_1-axis nor to the x_2-axis. In the case that the remaining nodes X_2, X_3 of the tetrahedra span an edge that is parallel to the x_2-axis the nodal functions φ_0 and φ_1 do not depend on x_2 and equation (3.18) is valid. Equation (3.19) follows then with (3.17). If the edge spanned by X_2 and X_3 is parallel to the x_1-axis a more detailed analysis is necessary. Therefore we rewrite (3.17) again by

$$(E_{0h}u - E_h u)|_T = (a_0 - a_1)\varphi_0 + a_1(\varphi_0 + \varphi_1).$$

A short computation shows that $\varphi_0 + \varphi_1 = 1 - x_3$ and, consequently,

$$\frac{\partial}{\partial x_2}(\varphi_0 + \varphi_1) = 0.$$

With $X_i = (X_{i,1}, X_{i,2}, X_{i,3})$, $i = 0, 1$ and $\Psi_{0,0} = \Psi_{0,1}$ one can write

$$|a_0 - a_1| = \left| \int_{\sigma_0} u(X_{0,1}, X_{0,2}, z) \Psi_{0,0}(z) \, \mathrm{d}z - \int_{\sigma_1} u(X_{1,1}, X_{1,2}, z) \Psi_{0,1}(z) \, \mathrm{d}z \right|$$

$$= \left| \int_{\sigma_0} \left[u(X_{0,1}, X_{0,2}, z) - u(X_{1,1}, X_{1,2}, z) \right] \Psi_{0,0}(z) \, \mathrm{d}z \right|.$$

The triangle inequality yields

$$|a_0 - a_1| \leq \left| \int_{\sigma_0} \left[u(X_{0,1}, X_{0,2}, z) - u(X_{1,1}, X_{0,2}, z) \right] \Psi_{0,0}(z) \, \mathrm{d}z \right|$$

$$+ \left| \int_{\sigma_0} \left[u(X_{1,1}, X_{0,2}, z) - u(X_{1,1}, X_{1,2}, z) \right] \Psi_{0,0}(z) \, \mathrm{d}z \right|$$

$$= \left| \int_{\sigma_0} \Psi_{0,0}(z) \int_{X_{1,1}}^{X_{0,1}} \frac{\partial}{\partial x_1} u(x, X_{0,2}, z) \, \mathrm{d}x \, \mathrm{d}z \right|$$

$$+ \left| \int_{\sigma_0} \Psi_{0,0}(z) \int_{X_{1,2}}^{X_{0,2}} \frac{\partial}{\partial x_2} u(X_{1,1}, y, z) \, \mathrm{d}y \, \mathrm{d}z \right|.$$

Now one can proceed as in the case of three nodes in Γ_B and arrives at

$$|a_0 - a_1| \leq ch_1^{-1} h_3^{-1} \sum_{\alpha \leq 1} h^\alpha \left[\left\| D^\alpha \left(\frac{\partial u}{\partial x_1} \right) \right\|_{L^1(S_T)} + \left\| D^\alpha \left(\frac{\partial u}{\partial x_2} \right) \right\|_{L^1(S_T)} \right].$$

Since

$$\left\| \frac{\partial}{\partial x_2} \varphi_0 \right\|_{L^q(T)} \leq ch_2^{-1} |T|^{1/q} \quad \text{and} \quad h_1 \leq h_2,$$

it follows as in the case of three nodes in Γ_B (comp. (3.16))

$$\left\| \frac{\partial}{\partial x_2} (E_{0h} - E_h) u \right\|_{L^q(T)} \leq c |T|^{1/q - 1/p} \sum_{|\alpha|=1} h^\alpha \left\| D^\alpha \left(\frac{\partial u}{\partial x_1} \right) \right\|_{L^p(S_T)} +$$

$$c |T|^{1/q - 1/p} \sum_{|\alpha|=1} h^\alpha \left\| D^\alpha \left(\frac{\partial u}{\partial x_2} \right) \right\|_{L^p(S_T)}.$$

It remains the case where only one node X_0 of T is contained in Γ_B. The difference of E_{0h} and E_h in T reduces to

$$(E_{0h} u - E_h u)_T = a_0 \varphi_0.$$

Since T is a tensor product element one has $\varphi_0 = \varphi_0(x_3)$ and consequently

$$\left\| \frac{\partial}{\partial x_2} (E_{0h} - E_h) u \right\|_{L^q(T)} = |a_0|^{1/q} \left\| \frac{\partial}{\partial x_2} \varphi_0 \right\|_{L^q(T)} = 0.$$

Summarizing all the cases we have shown

$$\left\|\frac{\partial}{\partial x_2}(E_{0h} - E_h)u\right\|_{L^q(T)} \le c\,|T|^{1/q-1/p} \sum_{|\alpha|=1} h^\alpha \|D^\alpha u\|_{W^{1,p}(S_T)}. \tag{3.20}$$

The proof for an estimate of the error in the x_1-derivative is analogous to the x_2-case and one gets

$$\left\|\frac{\partial}{\partial x_1}(E_{0h} - E_h)u\right\|_{L^q(T)} \le c\,|T|^{1/q-1/p} \sum_{|\alpha|=1} h^\alpha \|D^\alpha u\|_{W^{1,p}(S_T)}. \tag{3.21}$$

With (3.14), (3.20) and (3.21) the assertion is shown. $\qquad\square$

Theorem 3.5. *Consider an element T of a tensor product mesh and assume that (3.6) is fulfilled. Then the error estimate*

$$|u - E_{0h}u|_{W^{1,q}(T)} \le c\,|T|^{1/q-1/p} \sum_{|\alpha|=1} h_T^\alpha |D^\alpha u|_{W^{1,p}(S_T)} \tag{3.22}$$

holds for $p \in [1,\infty]$, q such that $W^{2,p}(T) \hookrightarrow W^{1,q}(T)$, $u \in W^{2,p}(S_T)$ and $u|_{\overline{T}\cap\Gamma_B} = 0$.

Proof. Inequality (3.22) follows with the triangle inequality from (3.7) and (3.12). $\qquad\square$

3.5 Local estimates in weighted Sobolev spaces

In order to get a global estimate for the interpolation error, it is useful to have an estimate where certain first derivatives of the interpolant are estimated against first derivatives of the function u. This additional stability estimate is necessary since we also consider functions $u \notin H^2(T)$ for elements T with $r_T = 0$. Thus we prove the following estimate for functions from weighted Sobolev spaces.

Lemma 3.6. *Consider a tensor product element T and assume that $h_{1,T} \sim h_{2,T} \le ch_{3,T}$. Let $p,q \in [1,\infty]$, $1 - 2/p < \beta < 2 - 2/p$ and $\beta \le 1$. Then one has for $u \in W^{1,p}(S_T) \cap V_\beta^{2,p}(S_T)$ and $u|_{\overline{T}\cap\Gamma_B} = 0$ the estimate*

$$|E_{0h}u|_{W^{1,q}(T)} \le c\,|T|^{1/q-1/p}h_{1,T}^{-\beta} \sum_{|\alpha|=1} h_T^\alpha \|D^\alpha v\|_{V_\beta^{1,p}(S_T)}. \tag{3.23}$$

For $u \in W^{1,p}(S_T) \cap W_\beta^{2,p}(S_T)$ the estimate

$$\|E_h u\|_{W^{1,q}(T)} \le c\,|T|^{1/q-1/p}h_{1,T}^{-\beta} \sum_{|\alpha|=1} h_T^\alpha \|D^\alpha v\|_{W_\beta^{1,p}(S_T)}. \tag{3.24}$$

is valid.

Proof. By the triangle inequality, Lemma 11 in [6] with $m = 1$ and (3.12) one has

$$
|E_{0h}u|_{W^{1,q}(T)} \leq |E_h u|_{W^{1,q}(T)} + |E_{0h}u - E_h u|_{W^{1,q}(T)}
$$
$$
\leq c\,|T|^{1/q-1} \sum_{|\alpha|\leq 1} h^\alpha |D^\alpha u|_{W^{1,1}(S_T)}. \tag{3.25}
$$

The step from $|\alpha| \leq 1$ to $|\alpha| = 1$ is analogous to the proof of Lemma 11 in [6]. One has just to substitute (6.13) in that proof by (3.25), and (3.23) is shown.

The second inequality can be proved in the following way. Since $\beta < 2 - 2/p$ one has $r^{-\beta} \in L^{p'}(S_T)$ with $1/p' = 1 - 1/p$ and one can write for $v \in W_\beta^{0,p}(S_T)$

$$
\|v\|_{L^1(S_T)} \leq \|r^{-\beta}\|_{L^{p'}(S_T)} \|r^\beta v\|_{L^p(S_T)}. \tag{3.26}
$$

Consider now two cylindrical sectors Z_1, Z_2 with radius $c_1 h_{1,T}$ and $c_2 h_{1,T}$ so that $Z_1 \subset S_T \subset Z_2$. Since $h_{1,T} \sim h_{2,T}$ we can conclude

$$
\left(\int_{Z_i} r^{-\beta p'}\right)^{1/p'} \sim \left(h_{3,T} \int_0^{c_i h_{1,T}} r^{-\beta p'+1}\right)^{1/p'} \sim \left(h_{3,T} h_{1,T}^{2-\beta p'}\right)^{1/p'} \sim \left(|S_T| \cdot h_{1,T}^{-\beta p'}\right)^{1/p'}
$$

for $i = 1, 2$. This results in the inequality

$$
\|r^{-\beta}\|_{L^{p'}(S_T)} \leq |S_T|^{1/p'} h_{1,T}^{-\beta}. \tag{3.27}
$$

The two inequalities (3.26) and (3.27) yield the embedding $W_\beta^{2,p}(S_T) \hookrightarrow W^{2,1}(S_T)$ and it follows $u \in W^{2,1}(S_T)$. Therefore one has from Theorem 10 in [6]

$$
\|E_h u\|_{W^{1,q}(T)} \leq c\,|T|^{1/q-1} \sum_{|\alpha|\leq 1} h_T^\alpha \|D^\alpha u\|_{W^{1,1}(S_T)}. \tag{3.28}
$$

Notice that the patch S_T defined in [6] is a subset of S_T as defined in Section 3.2. Now we continue from (3.28) with

$$
\|E_h u\|_{W^{1,q}(T)} \leq c\,|T|^{1/q-1} \sum_{|\alpha|\leq 1} \sum_{|t|\leq 1} h_T^\alpha \|D^{\alpha+t}u\|_{L^1(S_T)}
$$
$$
\leq c\,|T|^{1/q-1}|S_T|^{1-1/p} h_{1,T}^{-\beta} \sum_{|\alpha|\leq 1} \sum_{|t|\leq 1} h_T^\alpha \|r^\beta D^{\alpha+t}u\|_{L^p(S_T)}
$$
$$
\sim |T|^{1/q-1/p} h_{1,T}^{-\beta} \sum_{\alpha\leq 1} h_T^\alpha \|D^\alpha u\|_{W_\beta^{1,p}(S_T)}
$$

and the assertion (3.24) is shown. $\qquad\square$

In the following we prove an interpolation error estimate for functions in $W_\beta^{2,p}(T)$. This result is necessary for estimating the finite element error of pure Neumann problems.

Theorem 3.7. *Consider an element T of a tensor product mesh and assume that $h_{1,T} \sim h_{2,T} \leq c h_{3,T}$ is fulfilled. Then the error estimate*

$$\|u - E_h u\|_{W^{1,q}(T)} \leq c |T|^{1/q - 1/p} h_{1,T}^{-\beta} \sum_{|\alpha|=1} h_T^\alpha \|D^\alpha u\|_{W_\beta^{1,p}(S_T)} \tag{3.29}$$

holds for $p \in [1, \infty]$, $\beta \in (1 - 2/p, 1]$, q such that $W_\beta^{1,p}(T) \hookrightarrow L^q(T)$ and $u \in W_\beta^{2,p}(T)$.

Proof. From the triangle inequality we have for an arbitrary function $w \in W_\beta^{2,p}(S_T)$

$$\|u - E_h u\|_{W^{1,q}(T)} \leq \|u - w\|_{W^{1,q}(T)} + \|E_h(u - w)\|_{W^{1,q}(T)}. \tag{3.30}$$

For the first term in this inequality one can conclude with the embedding $W_\beta^{1,p}(S_T) \hookrightarrow L^q(S_T)$

$$\begin{aligned}
\|u - w\|_{W^{1,q}(T)} &\leq \|u - w\|_{W^{1,q}(S_T)} \\
&\leq c |T|^{1/q} \sum_{|t| \leq 1} h_T^{-t} \|D^t(\hat{u} - \hat{w})\|_{L^q(S_{\hat{T}})} \\
&\leq c |T|^{1/q} \sum_{|t| \leq 1} h_T^{-t} \|D^t(\hat{u} - \hat{w})\|_{W_\beta^{1,p}(S_{\hat{T}})} \\
&= c |T|^{1/q} \sum_{|t| \leq 1} \sum_{|\alpha| \leq 1} h_T^{-t} \|r^\beta D^{\alpha+t}(\hat{u} - \hat{w})\|_{L^p(S_{\hat{T}})}.
\end{aligned} \tag{3.31}$$

The application of (3.24) to $u - w$ yields

$$\begin{aligned}
\|E_h(u - w)\|_{W^{1,q}(T)} &\leq c |T|^{1/q - 1/p} h_{1,T}^{-\beta} \sum_{|\alpha| \leq 1} h_T^\alpha \|D^\alpha(u - w)\|_{W_\beta^{1,p}(S_T)} \\
&\leq c |T|^{1/q} \sum_{|\alpha| \leq 1} \sum_{|t| \leq 1} h_T^{-t} \|r^\beta D^{\alpha+t}(\hat{u} - \hat{w})\|_{L^p(S_{\hat{T}})}.
\end{aligned} \tag{3.32}$$

With (3.30), (3.31) and (3.32) one obtains

$$\begin{aligned}
\|u - E_h u\|_{W^{1,q}(T)} &\leq c |T|^{1/q} \sum_{|\alpha| \leq 1} \sum_{|t| \leq 1} h_T^{-t} \|r^\beta D^{\alpha+t}(\hat{u} - \hat{w})\|_{L^p(S_{\hat{T}})} \\
&= c |T|^{1/q} \sum_{|t| \leq 1} h_T^{-t} \|D^t(\hat{u} - \hat{w})\|_{W_\beta^{1,p}(S_{\hat{T}})}.
\end{aligned} \tag{3.33}$$

Now we specify w as the function of $\mathcal{P}_1(S_T)$, such that

$$\int_{S_{\hat{T}}} D^t(\hat{u} - \hat{w}) = 0 \qquad \forall t : |t| \leq 1,$$

and together with Lemma 2.17 one can continue from (3.33) with

$$\|u - E_h u\|_{W^{1,q}(T)} \leq c\,|T|^{1/q} \sum_{|t| \leq 1} h_T^{-t}|D^t \hat{u}|_{W_\beta^{1,p}(S_{\hat{T}})}$$

$$= c\,|T|^{1/q} \sum_{|t| \leq 1} \sum_{|\alpha|=1} h_T^{-t}\|r^\beta D^{\alpha+t}\hat{u}\|_{L^p(S_{\hat{T}})}$$

$$\leq c\,|T|^{1/q-1/p} h_{1,T}^{-\beta} \sum_{|t| \leq 1} \sum_{|\alpha|=1} h_T^\alpha\|r^\beta D^{\alpha+t}u\|_{L^p(S_T)}$$

$$= c\,|T|^{1/q-1/p} h_{1,T}^{-\beta} \sum_{|\alpha|=1} h_T^\alpha\|D^\alpha u\|_{W_\beta^{1,p}(S_T)}$$

and the assertion (3.29) is shown. $\qquad\qquad\qquad\qquad\qquad\qquad\qquad\square$

3.6 Extension to more general meshes

If we consider the special case $h_1 \sim h_2$, we can extend our results to more general meshes. Instead of tensor product elements we introduce as in [6] *elements of tensor product type*, that are defined by the transformation

$$\begin{pmatrix} x_1 \\ x_2 \\ x_3 \end{pmatrix} = \begin{pmatrix} B_T & 0 \\ 0 & \pm h_{3,T} \end{pmatrix} \begin{pmatrix} \hat{x}_1 \\ \hat{x}_2 \\ \hat{x}_3 \end{pmatrix} + b_T =: \hat{B} \begin{pmatrix} \hat{x}_1 \\ \hat{x}_2 \\ \hat{x}_3 \end{pmatrix} + b_T,$$

where $b_T \in \mathbb{R}^3$ and $B_T \in \mathbb{R}^{2 \times 2}$ with

$$|\det B_T| \sim h_{1,T}^2, \quad \|B_T\| \sim h_{1,T}, \quad \|B_T^{-1}\| \sim h_{1,T}^{-1}.$$

Additionally we introduce a coordinate system $\tilde{x}_1, \tilde{x}_2, \tilde{x}_3$ via the transformation

$$\begin{pmatrix} x_1 \\ x_2 \\ x_3 \end{pmatrix} = \begin{pmatrix} h_{1,T}^{-1} B_T & 0 \\ 0 & 1 \end{pmatrix} \begin{pmatrix} \tilde{x}_1 \\ \tilde{x}_2 \\ \tilde{x}_3 \end{pmatrix} =: \tilde{B} \begin{pmatrix} \tilde{x}_1 \\ \tilde{x}_2 \\ \tilde{x}_3 \end{pmatrix}.$$

This transformation maps T and S_T to $\tilde{T}$ and $\tilde{S}_T$. Since

$$\begin{pmatrix} \tilde{x}_1 \\ \tilde{x}_2 \\ \tilde{x}_3 \end{pmatrix} = \tilde{B}^{-1}\hat{B} \begin{pmatrix} \hat{x}_1 \\ \hat{x}_2 \\ \hat{x}_3 \end{pmatrix} + \tilde{B}^{-1} b_T = \begin{pmatrix} h_{1,T} & 0 & 0 \\ 0 & h_{1,T} & 0 \\ 0 & 0 & h_{3,T} \end{pmatrix} \begin{pmatrix} \hat{x}_1 \\ \hat{x}_2 \\ \hat{x}_3 \end{pmatrix} + \tilde{B}^{-1} b_T,$$

the mesh is a tensor product mesh in the coordinate system $\tilde{x}_1, \tilde{x}_2, \tilde{x}_3$. With $\tilde{S}_T = S_{\tilde{T}}$ it follows from

$$\det \tilde{B} \sim 1, \quad \|\tilde{B}\| \sim 1, \quad \|\tilde{B}^{-1}\| \sim 1$$

that our results extend to meshes of tensor product type.

Remark 3.8. The meshes defined in Subsection 2.3.2 are of tensor product type.

CHAPTER 4

Finite element error estimates for boundary value problems

One main ingredient of approximation error estimates in optimal control of partial differential equations are finite element error estimates for the state equation itself. Therefore we use this chapter to collect a couple of such results for several types of boundary value problems.

4.1 Scalar elliptic equations in polygonal domains

In this subsection we consider the boundary value problem

$$L_s y = f \quad \text{in } \Omega, \qquad y = 0 \quad \text{on } \Gamma = \partial\Omega, \tag{4.1}$$

over a bounded, polygonal domain $\Omega \subset \mathbb{R}^2$. The operator L_s is defined as

$$L_s y := -\nabla \cdot A(x)\nabla y + a_1(x) \cdot \nabla y + a_0(x)y$$

where $A \in C^\infty(\bar{\Omega}, \mathbb{R}^{2\times 2})$, $a_1 \in C^\infty(\bar{\Omega}, \mathbb{R}^2)$ and $a_0 \in C^\infty(\bar{\Omega})$. Furthermore, the coefficients are assumed to satisfy the conditions

$$m_0|\xi|^2 \leq \xi^T A(x)\xi \qquad \forall(\xi, x) \in \mathbb{R}^2 \times \bar{\Omega}, \ m_0 > 0$$

and

$$a_0(x) - \frac{1}{2}\nabla \cdot a_1(x) \geq 0 \qquad \forall x \in \Omega$$

ensuring ellipticity and coercivity, respectively. Additionally, we require A to be symmetric.

The variational formulation of this problem is given as

$$\text{Find } y \in V_0 : \qquad a_s(y,v) = (f,v)_{L^2(\Omega)} \qquad \forall v \in V_0 \qquad (4.2)$$

with the bilinear form $a_s : H^1(\Omega) \times H^1(\Omega) \to \mathbb{R}$,

$$a_s(y,v) := \int_\Omega A\nabla y \cdot \nabla v + (a_1 \cdot \nabla y)v + a_0 yv, \qquad (4.3)$$

and $V_0 := \{v \in H^1(\Omega) : v|_{\partial\Omega} = 0\}$. Notice, that existence and uniqueness of a solution of (4.2) is guaranteed by the Lax-Milgram lemma.

The singularities that are introduced by corners of the domain show local behavior. Therefore we reduce our considerations for simplicity to one corner with an interior angle ω located at the origin. More general situations can be reduced to this case by introducing suitable cut-off functions, see, e.g., [80].

4.1.1 Regularity

We consider the regularity properties of the solution of the boundary value problem (4.1). A short summary of relevant facts concerning the regularity of this boundary value problem is given in [15, Remark 1], which follows the outline by Sändig in [115]. In order to characterize the regularity of the solution y one considers the Dirichlet boundary value problem for the equation $L_{s,0}y = f$ in Ω, where $L_{s,0}$ is the principal part of the operator L_s with the coefficients evaluated at the corner point (here the origin). Then one particular eigenvalue of an operator pencil, that is obtained by an integral transformation of this modified problem, characterizes the regularity of y. It is worth mentioning, that in consequence the regularity is not influenced by the lower order terms with coefficients a_0 and a_1. In the following we denote the eigenvalue of interest by λ. In case of the Dirichlet problem for the Laplace operator in a two-dimensional domain and interior angle $\omega \in (\pi, 2\pi)$ at the reentrant corner, the eigenvalue λ is explicitly known, $\lambda = \pi/\omega$. This means $\lambda \in (1/2, 1)$. The more general case of the operator L_s is treated in [104, Chap. 5]. The linear coordinate transformation $y_1 = x_1 + d_1 x_2$, $y_2 = d_2 x_2$, with $d_1 = -a_{12}/a_{22}$ and $d_2 = \sqrt{a_{11}a_{22} - a_{12}^2}/a_{22}$, where a_{ij} $(i,j = 1,2)$ denote the coefficients of A, maps the differential operator $L_{s,0}$ to a multiple of the Laplace operator. Furthermore, the neighborhood of the corner, a sector with opening ω, is transformed in a circular sector with opening ω'. For $\omega \in (\pi, 2\pi)$ one has $\omega' \in (\pi, 2\pi)$ and therefore one can conclude that the quantity of interest, $\lambda = \pi/\omega'$, is also in the general case contained in the interval $(1/2, 1)$. The following lemma specifies the regularity of the solution y.

Lemma 4.1. *Let p and β be given real numbers with $p \in (1,\infty)$ and $\beta > 2 - \lambda - 2/p$, where λ is the particular eigenvalue of the operator pencil associated with L_s as described above. Moreover, let f be a function in $V_\beta^{0,p}(\Omega)$. Then the solution of the boundary value problem (4.2) belongs to $H_0^1(\Omega) \cap V_\beta^{2,p}(\Omega)$. Moreover, the inequality*

$$\|y\|_{V_\beta^{2,p}(\Omega)} \le c\|f\|_{V_\beta^{0,p}(\Omega)}$$

is valid.

Proof. With $\mathrm{Im}\lambda_- = -\lambda$ the assertion follows from Lemmata 1 and 2 of [115]. $\qquad\square$

The case $p = \infty$ is not included in the previous lemma. We use results of Maz'ya et al. [79, 90] to prove a regularity result in the space $V_\gamma^{2,\infty}(\Omega)$.

Lemma 4.2. *Let $\gamma \geq 2 - \lambda \geq 0$ and $f \in C^{0,\sigma}(\bar\Omega)$, $\sigma \in (0,1)$. Then the solution of the boundary value problem (4.2) belongs to the space $V_\gamma^{2,\infty}(\Omega)$, and the inequality*

$$\|y\|_{V_\gamma^{2,\infty}(\Omega)} \leq c\|f\|_{C^{0,\sigma}(\bar\Omega)}$$

is valid.

Proof. In [79], the weighted Hölder spaces $N_\beta^{l,\sigma}(\Omega)$ are introduced with the norm

$$\|y\|_{N_\beta^{l,\sigma}(\Omega)} = \sup_{x\in\Omega} \sum_{|\alpha|\leq l} r^{\beta-l-\sigma+|\alpha|}|D^\alpha y| + \sum_{|\alpha|=l} \sup_{x,x'\in\Omega} \frac{\left|\left(r^\beta D^\alpha y\right)(x) - \left(r^\beta D^\alpha y\right)(x')\right|}{|x-x'|^\sigma}.$$

In section 8.7.1 of [79] it is shown that for $-\lambda \leq l + \sigma - \beta \leq \lambda$ the regularity result

$$\|y\|_{N_\beta^{l,\sigma}(\Omega)} \leq c\|f\|_{N_\beta^{l-2,\sigma}(\Omega)}$$

for the solution y of (4.2) holds. This means that in the case of $l = 2$ one can conclude for $\gamma := \beta - \sigma \geq 2 - \lambda$

$$r^{\gamma-2+|\alpha|}|D^\alpha y| \leq c\|f\|_{N_{\gamma+\sigma}^{0,\sigma}(\Omega)} \quad \forall \alpha : |\alpha| \leq 2,$$

and therefore

$$\|y\|_{V_\gamma^{2,\infty}(\Omega)} \leq c\|f\|_{N_\beta^{0,\sigma}(\Omega)}.$$

According to [90, section 5] the $N_\beta^{l,\sigma}(\Omega)$-norm is equivalent to

$$\sup_{\substack{x,x'\in\Omega \\ 2|x-x'|\leq\min\{|x|,|x'|\}}} r(x)^\beta \sum_{|\alpha|=l} \frac{|D^\alpha y(x) - D^\alpha y(x')|}{|x-x'|^\sigma} + \sup_{x\in\Omega} r(x)^{\beta-l-\sigma}|y(x)|.$$

If one sets $l = 0$, this implies with $\gamma = \beta - \sigma \geq 0$ the embedding $C^{0,\sigma}(\bar\Omega) \hookrightarrow N_\beta^{0,\sigma}(\Omega)$ and the assertion is shown. $\qquad\square$

4.1.2 Finite element error estimates

Let us first introduce the finite element solution y_h of the boundary value problem (4.2). To this end, we set V_{0h} as the space of all piecewise linear and globally continuous functions in Ω, that vanish on the boundary $\partial\Omega$,

$$V_{0h} := \{v_h \in C(\bar\Omega) : v_h|_T \in \mathcal{P}_1 \text{ for all } T \in T_h \text{ and } v_h = 0 \text{ on } \partial\Omega\}.$$

Then the discretized problem reads as

$$\text{Find } y_h \in V_{0h}: \qquad a_s(y_h, v_h) = (f, v_h)_{L^2(\Omega)} \qquad \forall v_h \in V_{0h}. \qquad (4.4)$$

The Lax-Milgram lemma is also applicable in the discrete case and therefore it exists a unique solution y_h of (4.4). We recall the well-known result for the finite element error in the H^1- and L^2-Norm, see [22, 104, 109].

Theorem 4.3. *Let y and y_h be the solution of (4.2) and (4.4), respectively. On a mesh of type (2.12) with grading parameter $\mu < \lambda$ the estimate*

$$\|y - y_h\|_{L^2(\Omega)} + h\|y - y_h\|_{H^1(\Omega)} \leq ch^2 |y|_{V_\beta^{2,2}(\Omega)} \leq ch^2 \|f\|_{L^2(\Omega)}$$

is valid for $\beta > 1 - \lambda$.

The remainder of this subsection is devoted to an L^∞-estimate of the finite element error. As already mentioned in Chapter 1 there were many results on this topic published in the 1970's. But all of them are not suitable for our setting due to a restriction on quasi-uniform meshes, strong regularity assumptions on the solution or the domain or missing exact regularity assumptions on the right-hand side. Therefore we extend these results in the following theorem. This was originally published in [14]. A mistake in [14, Lemma 2.13] is corrected in [5].

Theorem 4.4. *Let y be the solution of the boundary value problem (4.2) with a right-hand side $f \in C^{0,\sigma}(\bar{\Omega})$, $\sigma \in (0,1)$. The finite element error can be estimated by*

$$\|y - y_h\|_{L^\infty(\Omega)} \leq ch^2 |\ln h|^{3/2} \|f\|_{C^{0,\sigma}(\bar{\Omega})}$$

on finite element meshes of type (2.12) with grading parameter $\mu < \lambda/2$.

Remark 4.5. A comparison of Theorem 4.3 and Theorem 4.4 shows that in order to achieve the proposed approximation rate in $L^\infty(\Omega)$ a stronger mesh grading is necessary than in $L^2(\Omega)$. A mesh is graded if $\mu < 1$. This means that the condition $\mu < \frac{\lambda}{2}$ yields a graded mesh not only in the case of a reentrant corner but also for corners with interior angle beyond a critical angle $\omega_0 < \pi$ for which $\lambda = 2$. For the Laplace operator, where the eigenvalue λ is explicitly known as $\lambda = \pi/\omega$ (comp. also the discussion in Subsection 4.1.1), one has $\omega_0 = \pi/2$. Consequently, already convex domains require mesh adaption.

The remainder of this section concerns the proof of Theorem 4.4. For the error analysis we cover the domain Ω with sectors Ω_{R_i}. The sectors are centered in the corners of Ω and if necessary also in other points on the boundary $\partial\Omega$ such that $\partial\Omega \subset \bigcup \bar{\Omega}_{R_i}$. The radii r_i of Ω_{R_i} are chosen small enough such that there exist circles/sectors $\tilde{\Omega}_{R_i} \subset \Omega$ and $\hat{\Omega}_{R_i} \subset \Omega$ with the same center as Ω_{R_i} but with radii $1.25 r_i$ and $1.5 r_i$. We define $\Omega_0 := \Omega \backslash \bigcup \bar{\Omega}_{R_i}$. The domains Ω_{R_i} are chosen such that Ω_0 is an interior subset of Ω. In Fig. 4.1 an example of such a partition is illustrated. The dotted and the dashed line in this figure show domains $\tilde{\Omega}_{R_i}$ and $\hat{\Omega}_{R_i}$, respectively. In the following we concentrate on one corner i_0 with an interior angle larger than the critical value ω_0 and assume that it is

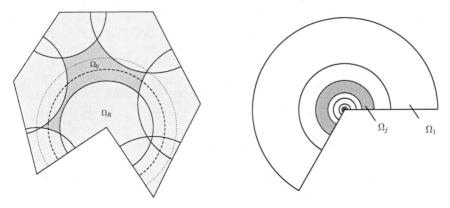

Figure 4.1: Partition of Ω in domains Ω_{R_i} (left) and of Ω_R in domains Ω_j (right)

located at the origin. We further assume that the sector $\Omega_{R_{i_0}}$ is centered at this corner and denote it by Ω_R, i.e. we suppress the index i_0. For the sake of simplicity we assume without loss of generality that $r_{i_0} = 1$. We split Ω_R in subsets Ω_j,

$$\Omega_R = \bigcup_{j=1}^{I} \Omega_j$$

where $\Omega_I = \{x : |x| \leq d_I\}$ and $\Omega_j = \{x : d_{j+1} \leq |x| \leq d_j\}$ for $j = 1, \ldots, I-1$. We set the radii d_j to $d_j = 2^{-j}$ $(j = 1, \ldots, I)$. The largest index I is chosen such that

$$d_I = ch^{2/\lambda}, \tag{4.5}$$

which means $I \sim \log \frac{1}{h}$. Further, we introduce the extended domains

$$\Omega'_j = \Omega_{j-1} \cup \Omega_j \cup \Omega_{j+1}$$

with the obvious modification for $j = 0$ and $j = I$ and in the same way, we define

$$\Omega''_j = \Omega'_{j-1} \cup \Omega'_j \cup \Omega'_{j+1}.$$

The subdomain meshsizes in Ω_j, $j = 1, \ldots, I$ are denoted by

$$h_j = \max_{T \in \Omega_j} h_T.$$

See also Fig. 4.1 for an illustration. Notice that the elements in Ω_j, $j = 1, \ldots, I-1$ are of comparable mean size, but the elements of Ω_I are not, see Lemma 4.6.

In the following we prove a couple of lemmas that contain auxiliary results which we need for the proof of Theorem 4.4. Our roadmap is a follows. First of all, we prove two results concerning the subdomain meshsizes h_j (Lemma 4.6 and 4.7). Later they will be useful for applying a recursive argumentation. Another necessary tool will be the estimate of the L^∞-norm of functions from V_h in a subdomain against the H^1-norm in the corresponding extended subdomain (Lemma 4.9). Then we can go ahead with an estimate for the local error $\|y - y_h\|_{L^\infty(\Omega_J)}$ subject to the L^2- and H^1-error in the extended domain Ω'_J (Lemma 4.10). The H^1-part can be estimated with respect to the L^2-error in Ω'_J (Lemma 4.12). Finally Lemma 4.13 gives an upper bound for $\|\tilde{y} - \tilde{y}_h\|_{L^2(\Omega'_J)}$, where $\tilde{y} = \eta y$ with a suitable cut-off function η. With this auxiliary results at hand the proof of Theorem 4.4 can be completed.

Lemma 4.6. *For the mesh introduced in (2.12) one has in Ω_j, $j = 1, \ldots, I - 1$ a family of quasi-uniform meshes with local mesh parameter*

$$2^{\mu-1}c_1 h d_j^{1-\mu} \leq h_j \leq c_2 h d_j^{1-\mu} \quad j = 0 \ldots I - 1 \tag{4.6}$$

with constants c_1 and c_2 from (2.12). For elements $T \subset \Omega_I$ the element size h_T satisfies

$$c_1 h^{1/\mu} \leq h_T \leq c_2 h d_I^{1-\mu} \tag{4.7}$$

with constants c_1 and c_2 from (2.12).

Proof. For $T \subset \Omega_j$ ($j \neq I$) one has $d_{j+1} \leq r_T \leq d_j$ and, therefore, $c_1 h d_{j+1}^{1-\mu} \leq h_T \leq c_2 h d_j^{1-\mu}$. Since $d_{j+1} = \frac{1}{2}d_j$ assertion (4.6) follows. For an element $T \subset \Omega_I$ one has $0 \leq r_T \leq d_I$, and therefore according to (2.12) the inequality $c_1 h^{1/\mu} \leq h_T \leq c_2 h d_I^{1-\mu}$ is valid, what is (4.7). $\qquad\square$

Lemma 4.7. *For every fixed $c_0 < 1$ and $\alpha > 0$ there exists $h_0 < 1$ such that*

$$h_j d_j^{-1} |\ln h|^\alpha \leq c_0$$

for $h < h_0$ and $j = 1, \ldots, I$.

Proof. With Lemma 4.6 it follows

$$h_j d_j^{-1} |\ln h|^\alpha \leq c h d_j^{1-\mu} d_j^{-1} |\ln h|^\alpha = c h d_j^{-\mu} |\ln h|^\alpha$$

$$\leq c h d_I^{-\mu} |\ln h|^\alpha \leq c h \left(h^{1/\mu - \varepsilon} \right)^{-\mu} |\ln h|^\alpha$$

$$\leq c h^{\varepsilon\mu} |\ln h|^\alpha .$$

Since this last value tends to zero as h tends to zero the assertion follows. $\qquad\square$

We recall a lemma from [129, Corollary 2.1] and use it for the proof of a variant of Sobolev's inequality.

Lemma 4.8. *Let $G \subset \mathbb{R}^d$ be a bounded domain with Lipschitz boundary and $t > 1$. If $qs > d$, $\lambda = \min\{1, s - d/q\}$, then there exists a constant c such that*

$$\|u\|_{C(\bar{G})} \leq c \left(|\ln \varepsilon|^{1-1/d} \|u\|_{W^{d/t,t}(G)} + \varepsilon^\lambda \|u\|_{W^{s,q}(G)} \right)$$

holds for any $\varepsilon > 0$ and any $u \in W^{d/t,t}(G) \cap W^{s,q}(G)$.

Lemma 4.9. *For every* $v_h \in V_h$ *and every* $J \in \{1, \ldots, I\}$ *the estimates*

$$\|v_h\|_{L^\infty(\Omega_J)} \leq c \, |\ln h_J|^{1/2} \, \|v_h\|_{H^1(\Omega'_J)} \quad \forall J \in \{1, \ldots, I\} \tag{4.8}$$

$$\|v_h\|_{L^\infty(\Omega)} \leq c \, |\ln h|^{1/2} \, \|v_h\|_{H^1(\Omega)} \tag{4.9}$$

are valid.

Proof. The following proof is adapted from the proof of [129, Theorem 3.4]. The modification is necessary in order to obtain constants independent of the domains Ω_J. Moreover, the union of all triangles T with $\bar{T} \cap \Omega_J \neq \emptyset$ is in general a proper superset of Ω_J. By using the coordinate transformation $x \mapsto d_J \hat{x}$ we map Ω_J for all $J = 1, \ldots, I-1$ to the same domain $\hat{\Omega}$. Then we apply Lemma 4.8 with $s = 1$, $t = d = 2$, $q = \infty$, and $\varepsilon = h_J$ for $v_h \in V_h$ and use Poincaré type inequalities on $\hat{\Omega}$ where we employ that $\hat{v}_h$ satisfies Dirichlet boundary conditions on part of $\partial\hat{\Omega}$. So we conclude with the help of standard scaling arguments

$$\begin{aligned}
\|v_h\|_{L^\infty(\Omega_J)} &= \|\hat{v}_h\|_{L^\infty(\hat{\Omega})} \\
&\leq c \left(|\ln h_J|^{1/2} \|\hat{v}_h\|_{H^1(\hat{\Omega})} + h_J \|\hat{v}_h\|_{W^{1,\infty}(\hat{\Omega})} \right) \\
&\leq c \left(|\ln h_J|^{1/2} \left\|\hat{\nabla}\hat{v}_h\right\|_{L^2(\hat{\Omega})} + h_J \|\hat{\nabla}\hat{v}_h\|_{L^\infty(\hat{\Omega})} \right) \\
&\leq c \left(|\ln h_J|^{1/2} \|\nabla v_h\|_{L^2(\Omega_J)} + h_J d_J \|\nabla v_h\|_{L^\infty(\Omega_J)} \right).
\end{aligned}$$

The application of the inverse inequality

$$h_J \|\nabla v_h\|_{L^\infty(\Omega_J)} \leq \|\nabla v_h\|_{L^2(\Omega'_J)}$$

as well as the fact $d_J \leq 1$ yield the assertion (4.8). The proof is the same for $J = I$, only the reference domain $\hat{\Omega}$ is another one. Recalling that $h_T \geq c h^{1/\mu}$ for any T, and therefore $|\ln h_T| \leq c |\ln h|$ we obtain (4.9) without the need of the scaling argument. $\square$

Lemma 4.10. *For* $y \in V^{2,2}_\beta(\Omega_J) \cap V^{2,\infty}_\gamma(\Omega_J)$ *with* $\beta = 1 - \lambda + \delta$, $\gamma = 2 - \lambda$, $\mu = \frac{\lambda}{2} - \delta'$, $\delta, \delta' > 0$ *the estimates*

$$\|y - y_h\|_{L^\infty(\Omega_J)} \leq c \left(h^2 \, |\ln h| \, |y|_{V^{2,\infty}_\gamma(\Omega''_J)} + d_J^{-1} \|y - y_h\|_{L^2(\Omega'_J)} \right) \text{ for } J < I-2, \tag{4.10}$$

$$\|y - y_h\|_{L^\infty(\Omega_J)} \leq c \left(|\ln h|^{1/2} \, h^2 \|y\|_{V^{2,\infty}_\gamma(\Omega'_J)} + |\ln h|^{1/2} \, \|y - y_h\|_{H^1(\Omega'_J)} \right)$$
$$\text{for } J \geq I-2 \tag{4.11}$$

are valid.

Proof. Let us first consider the case $J < I-2$, where one is away from the corner. We use the estimate

$$\|y - y_h\|_{L^\infty(\Omega_J)} \leq c \left(|\ln h| \min_{\chi \subset V_h} \|y - \chi\|_{L^\infty(\Omega'_J)} + d_J^{-1} \|y - y_h\|_{L^2(\Omega'_J)} \right). \tag{4.12}$$

This result follows from Theorem 5.1 in [116] as shown in the proof of Corollary 5.1 of that paper, where the authors have already inserted an interpolation error estimate. If one chooses $l = 0$, $N = 2$, $p = 0$, and $q = 2$ in that corollary, inequality (4.12) follows from writing $y - y_h$ as $y - \chi - y_h + \chi$. For an application of this result to the domains Ω_J we refer also to [43, Example 10.1]. If one assumes for the Lagrange interpolant I_h that $y_h - I_h y$ admits its maximum in Ω'_J in $x_0 \in \bar{T}_* \subset \Omega''_J$, one can conclude

$$\|y - I_h y\|_{L^\infty(\Omega'_J)} = \|y - I_h y\|_{L^\infty(T_*)}$$
$$\leq c h_{T_*}^2 |y|_{W^{2,\infty}(T_*)}$$
$$\leq c h_J^2 |y|_{W^{2,\infty}(\Omega''_J)}$$
$$\sim h^2 d_J^{2-2\mu} |y|_{W^{2,\infty}(\Omega''_J)}$$
$$\sim h^2 d_J^{2-2\mu-\gamma} |y|_{V_\gamma^{2,\infty}(\Omega''_J)}.$$

Since $2 - 2\mu - \gamma = \lambda - 2(\frac{\lambda}{2} - \delta') = 2\delta' > 0$ this yields

$$\|y - I_h y\|_{L^\infty(\Omega'_J)} \leq c h^2 |y|_{V_\gamma^{2,\infty}(\Omega''_J)}.$$

Then the assertion (4.10) follows from inequality (3.31).

Let us now consider the case of $J = I, I - 1, I - 2$. With the triangle inequality it follows

$$\|y - y_h\|_{L^\infty(\Omega_J)} \leq \|y\|_{L^\infty(\Omega_J)} + \|y_h\|_{L^\infty(\Omega_J)}. \tag{4.13}$$

We estimate the two terms separately. From the embedding (2.2) one can conclude

$$\|y\|_{L^\infty(\Omega_J)} \leq c d_I^{2-\gamma} \|y\|_{V_\gamma^{2,\infty}(\Omega_J)} = c d_I^\lambda \|y\|_{V_\gamma^{2,\infty}(\Omega_J)} = c h^2 \|y\|_{V_\gamma^{2,\infty}(\Omega_J)} \tag{4.14}$$

where we used equation (4.5) in the last step. In order to estimate the second term of (4.13) we use Lemma 4.9 and get the inequality

$$\|y_h\|_{L^\infty(\Omega_J)} \leq c |\ln h_J|^{1/2} \|y_h\|_{H^1(\Omega'_J)} \leq c |\ln h|^{1/2} \|y_h\|_{H^1(\Omega'_J)}. \tag{4.15}$$

where we have used $h_J \leq c h d_I^{1-\mu} = c h^{1+2/\lambda(1-\mu)} \leq c h^{2/\lambda}$ in the last step. We use again the triangle inequality to estimate

$$\|y_h\|_{H^1(\Omega'_J)} \leq \|y\|_{H^1(\Omega'_J)} + \|y - y_h\|_{H^1(\Omega'_J)}. \tag{4.16}$$

In order to estimate the first part of the right-hand side of this inequality we continue with $\alpha = \gamma - 1 = 1 - \lambda$

$$\|y\|_{H^1(\Omega'_J)} \sim \left\|r^{-\alpha} r^\alpha \nabla y\right\|_{L^2(\Omega'_J)} + \left\|r^{1-\alpha} r^{\alpha-1} y\right\|_{L^2(\Omega'_J)}$$
$$\leq \left\|r^{-\alpha}\right\|_{L^2(\Omega'_J)} \left\|r^\alpha \nabla y\right\|_{L^\infty(\Omega'_J)} + \left\|r^{1-\alpha}\right\|_{L^2(\Omega'_J)} \left\|r^{\alpha-1} y\right\|_{L^\infty(\Omega'_J)}$$
$$\leq c d_I^{1-\alpha} \|y\|_{V_\alpha^{1,\infty}(\Omega'_J)}$$
$$\leq c d_I^\lambda \|y\|_{V_{\alpha+1}^{2,\infty}(\Omega'_J)}$$
$$\leq c h^2 \|y\|_{V_\gamma^{2,\infty}(\Omega'_J)},$$

since $d_I = ch^{2/\lambda}$, see (4.5). With this estimate one has from (4.15) and (4.16)

$$\|y_h\|_{L^\infty(\Omega_J)} \le c \left(|\ln h|^{1/2} h^2 \|y\|_{V_\gamma^{2,\infty}(\Omega_J')} + |\ln h|^{1/2} \|y - y_h\|_{H^1(\Omega_J')} \right)$$

which yields together with (4.13) and (4.14) the desired result (4.11). □

Lemma 4.11. *The estimate*

$$\|y - y_h\|_{H^1(\Omega_J)} \le c \left(\|y - I_h y\|_{H^1(\Omega_J')} + d_J^{-1} \|y - I_h y\|_{L^2(\Omega_J')} + d_J^{-1} \|y - y_h\|_{L^2(\Omega_J')} \right)$$

is valid for $J = 1, \ldots, I$.

Proof. The assertion follows from Lemma 7.2 of [117] by setting $D_1 = \Omega_J$, $D = \Omega_J'$, and $p = 0$ in that lemma. It follows from Lemma 4.6 and the explanations in Example 4 of section 9 in [117] that the result is applicable with our finite element space. Notice that the proof is only given for $L = -\Delta$. For an extension to general elliptic operators the proof has to be modified at two points, where the bilinear form explicitly steps in. After equation (7.7) in that proof one has to substitute the estimate of $\|v_h\|_{1,D_1}^2$ by

$$\|v_h\|_{1,D_1}^2 \le \|\omega v_h\|_{1,D}^2$$

$$\le c a_s \left(v_h, \omega^2 v_h \right) + c \int_\Omega (\nabla \cdot (A \nabla w) \omega) \omega v_h + v_h \left(2 A \nabla \omega \cdot \nabla(\omega v_h) \right) + \int_{D \setminus D_1} v_h^2 \omega a_1 \cdot \nabla \omega.$$

Notice, that in [117] A is the bilinear form while in our setting a_s is the bilinear form and A the coefficient matrix in the operator L. Since $A \subset W^{1,\infty}(\Omega, \mathbb{R}^{2,2})$ and $\|a_1 \cdot \nabla \omega\|_{L^\infty(\Omega)} \le C$ it follows that

$$\|\omega v_h\|_{1,D_1}^2 \le Ch \|v_h\|_{1,D}^2 + C \|v_h\|_{0,D \setminus D_1} \|\omega v_h\|_{1,D_1}^2$$

and therefore equation (7.8) in [117]. The second point is after expression (7.10), where the equation for $(\omega v_h, \varphi)$ has to be substituted by

$$(\omega v_h, \varphi) = a_s(\omega v_h, \psi) = a_s(v_h, \omega \psi) + \int_D v_h \left(\psi \nabla \cdot (A \nabla \omega) + 2 A \nabla \omega \cdot \nabla \psi + \psi a_1 \cdot \nabla \omega \right).$$

Again it follows from the fact $A \in W^{1,\infty}(\Omega, \mathbb{R}^{2,2})$ and $\|a_1 \cdot \nabla \omega\|_{L^\infty(\Omega)} \le C$ that the argumentation can be completed as for the Laplace operator. □

Lemma 4.12. *For $y \in V_\beta^{2,2}(\Omega_J) \cap V_\gamma^{2,\infty}(\Omega_J)$, $\beta = 1 - \lambda + \delta$, $\gamma = 2 - \lambda$, and $\mu = \frac{\lambda}{2} - \delta'$, $\delta, \delta' > 0$, the estimates*

$$\|y - y_h\|_{H^1(\Omega_J)} \le c \left(h d_J^{1-\mu-\beta} |y|_{V_\beta^{2,2}(\Omega_J'')} + h^{1+2\delta'/\lambda} d_J^{2-\gamma-\mu} |y|_{V_\gamma^{2,\infty}(\Omega_J'')} + \right.$$

$$\left. d_J^{-1} \|y - y_h\|_{L^2(\Omega_J')} \right) \quad \text{for } J < I - 2 \quad (4.17)$$

$$\|y - y_h\|_{H^1(\Omega_J)} \le c \left(h^2 \|y\|_{V_\gamma^{2,\infty}(\Omega_J'')} + d_J^{-1} \|y - y_h\|_{L^2(\Omega_J')} \right) \quad \text{for } J \ge I - 2 \quad (4.18)$$

are valid.

Proof. From Lemma 4.11 we have

$$\|y - y_h\|_{H^1(\Omega_J)} \le c \left(\|y - I_h y\|_{H^1(\Omega_J')} + d_J^{-1}\|y - I_h y\|_{L^2(\Omega_J')} + d_J^{-1}\|y - y_h\|_{L^2(\Omega_J')} \right)$$

$$\le c \left(\|y - I_h y\|_{H^1(\Omega_J')} + \|y - I_h y\|_{L^\infty(\Omega_J')} + d_J^{-1}\|y - y_h\|_{L^2(\Omega_J')} \right), \quad (4.19)$$

where we have used $\|y - I_h y\|_{L^2(\Omega_J')} \le c|\Omega_J'|^{1/2}\|y - I_h y\|_{L^\infty(\Omega_J')}$ and $|\Omega_J'| \sim d_J^2$. In the case $J = 0, \ldots, I - 3$ one has $y \in H^2(\Omega_J'')$ and it follows for $T \in \Omega_{J,h}'$

$$\|y - I_h y\|_{H^1(T)} \le ch_J|y|_{H^2(T)} \le ch_J r_T^{-\beta}\|r^\beta \nabla^2 y\|_{L^2(T)}$$

$$\le chd_J^{1-\mu-\beta}|y|_{V_\beta^{2,2}(T)},$$

where we used $r_T \sim d_J$ for $T \in \Omega_{J,h}'$ and $h_J = hd_J^{1-\mu}$. This yields

$$\|y - I_h y\|_{H^1(\Omega_J')} \le hd_J^{1-\mu-\beta}|y|_{V_\beta^{2,2}(\Omega_J'')}. \quad (4.20)$$

Like in the proof of Lemma 4.10 the second term on the right-hand side of inequality (4.19) can be estimated by

$$\|y - I_h y\|_{L^\infty(\Omega_J')} \sim h^2 d_J^{2-2\mu-\gamma}|y|_{V_\gamma^{2,\infty}(\Omega_J'')}$$

$$\le chd_J^{-\mu} \cdot hd_J^{2-\gamma-\mu}|y|_{V_\gamma^{2,\infty}(\Omega_J'')}$$

$$\le ch^{1+2\delta'/\lambda}d_J^{2-\gamma-\mu}|y|_{V_\gamma^{2,\infty}(\Omega_J'')}$$

since $hd_J^{-\mu} \le hd_I^{-\mu} = chh^{-\frac{2}{\lambda}(\frac{\lambda}{2}-\delta')} = ch^{2\delta'/\lambda}$. This last estimate yields together with estimate (4.20) and inequality (4.19) assertion (4.17).

In the case $J = I, I - 1, I - 2$ we distinguish again between elements with $r_T = 0$ and $r_T > 0$. For elements T with $r_T = 0$ we write

$$\|y - I_h y\|_{H^1(T)} \le \|y\|_{H^1(T)} + \|I_h y\|_{H^1(T)}. \quad (4.21)$$

For the first term one has, like in the proof of Lemma 4.10 after (4.16),

$$\|y\|_{H^1(T)} \le ch^2\|y\|_{V_\gamma^{2,\infty}(T)}.$$

In order to estimate the second term, we conclude with the inverse inequality, the estimate $\|I_h y\|_{L^\infty(\Omega_J)} \le \|y\|_{L^\infty(\Omega_J)}$ and the embedding (2.2)

$$\|I_h y\|_{H^1(T)} \le ch_J^{-1}\|I_h y\|_{L^2(T)} \le c\|y\|_{L^\infty(T)}$$

$$\le ch_T^{2-\gamma}\|y\|_{V_\gamma^{2,\infty}(T)} \le ch^2\|y\|_{V_\gamma^{2,\infty}(T)} \quad (4.22)$$

since $h_T \sim h^{1/\mu}$ and $2 - \gamma = \lambda > 2\mu$. The inequalities (4.21)–(4.22) yield

$$\|y - I_h y\|_{H^1(T)} \le ch^2\|y\|_{V_\gamma^{2,\infty}(T)}.$$

For elements T with $r_T > 0$ and $T \cap \Omega_J \neq \emptyset$ we can write

$$\|y - I_h y\|_{H^1(T)} \leq c h_T |y|_{H^2(T)} \leq h |y|_{V^{2,2}_{1-\mu}(T)}$$

which yields

$$\left(\sum_{T : r_T > 0 \wedge T \cap \Omega_J \neq \emptyset} \|y - I_h y\|^2_{H^1(T)} \right)^{1/2} \leq c h |y|_{V^{2,2}_{1-\mu}(\Omega'_J)}.$$

With the Hölder inequality one can conclude

$$|y|_{V^{2,2}_{1-\mu}(\Omega'_J)} = \left(\int_{\Omega'_J} r^{2(1-\mu)} |D^2 y|^2 \right)^{1/2} = \left(\int_{\Omega'_J} r^{2\lambda - 2\mu - 2} r^{2(2-\lambda)} |D^2 y|^2 \right)^{1/2}$$

$$\leq |y|_{V^{2,\infty}_{2-\lambda}(\Omega'_J)} \left(\int_{\Omega'_J} r^{2\lambda - 2\mu - 2} \right)^{1/2} = c d_I^{\lambda - \mu} |y|_{V^{2,\infty}_{2-\lambda}(\Omega'_J)} \leq c h |y|_{V^{2,\infty}_{2-\lambda}(\Omega'_J)}.$$

Sticking everything together yields

$$\|y - I_h y\|_{H^1(\Omega_J)} \leq c h^2 |y|_{V^{2,\infty}_{\gamma}(\Omega'_J)}. \tag{4.23}$$

It remains to estimate the second term in (4.19) in the case $J = I, I-1, I-2$. We assume that the maximum of $y - I_h y$ in Ω'_J is attained in $\bar{T}_* \subset \bar{\Omega}''_J$. It follows

$$\|y - I_h y\|_{L^\infty(\Omega'_J)} \leq \|y - I_h y\|_{L^\infty(T_*)} \leq c \|y\|_{L^\infty(T_*)}.$$

and with the same argumentation as in the lines before inequality (4.14) one can conclude

$$\|y - I_h y\|_{L^\infty(\Omega'_J)} \leq c h^2 \|y\|_{V^{2,\infty}_{\gamma}(\Omega''_J)}.$$

This estimate yields together with (4.19) and (4.23) the desired inequality (4.18). □

We define the function

$$\tilde{y} = \eta y \tag{4.24}$$

where $\eta = \eta(r)$ is a smooth cut-off function with $\eta \equiv 1$ in $\tilde{\Omega}_R$ and $\eta \equiv 0$ in $\Omega \backslash \hat{\Omega}_R$. The function $\tilde{y}$ can be seen as solution of a mixed boundary value problem with right-hand side $L\tilde{y}$ and Dirichlet conditions on $\partial \Omega_R \cap \partial \Omega$ and Neumann conditions on $\partial \Omega_R \backslash (\partial \Omega_R \cap \partial \Omega)$. The Ritz projection of $\tilde{y}$ is denoted by $\tilde{y}_h$, i.e.,

$$a(\tilde{y} - \tilde{y}_h, \chi) = 0 \quad \forall \chi \in V_h. \tag{4.25}$$

Notice that Lemmas 4.10 − 4.12 are also valid for $\tilde{y}$ and $\tilde{y}_h$ since we used only Galerkin orthogonality and interpolation error estimates in the proofs.

Lemma 4.13. *Under the conditions of Lemma 4.12 and $2\delta' > \delta$ the inequality*

$$d_J^{-1}\|y - y_h\|_{L^2(\Omega_J')} \le ch^2\,|\ln h|\,\Big(\|y\|_{V_\beta^{2,2}(\Omega)} + \|y\|_{V_\gamma^{2,\infty}(\Omega)}\Big)$$

is valid for $J = 1, \dots, I$.

Proof. For this proof we introduce the abbreviation $\tilde{e} := \tilde{y} - \tilde{y}_h$ with $\tilde{y}$ and $\tilde{y}_h$ from (4.24) and (4.25), respecitvely. One has the equality

$$\|\tilde{e}\|_{L^2(\Omega_R)} = \sup_{\substack{\varphi \in C_0^\infty(\Omega_R) \\ \|\varphi\|_{L^2(\Omega_R)}=1}} (\tilde{e}, \varphi). \tag{4.26}$$

For every such function φ we consider the boundary value problem

$$\begin{aligned}
Lv = -\nabla \cdot A(x)\nabla v - a_1(x) \cdot \nabla v + a_0(x)v = (|x| + d_I)^{-1}\varphi &\quad \text{in } \Omega_R, \\
v = 0 &\quad \text{on } \partial\Omega_R \cap \partial\Omega, \\
(A\nabla v) \cdot n = 0 &\quad \text{on } \partial\Omega_R \backslash(\partial\Omega_R \cap \partial\Omega),
\end{aligned}$$

with its weak formulation

$$a_{s,\Omega_R}(w, v) = ((|x| + d_I)^{-1}\varphi, w)_{\Omega_R} \quad \forall w \in \{w \in H^1(\Omega_R) : w = 0 \text{ on } \partial\Omega_R \cap \partial\Omega\}.$$

Then one can conclude

$$\begin{aligned}
((|x| + d_I)^{-1}\tilde{e}, \varphi)_{\Omega_R} = (\tilde{e}, (|x| + d_I)^{-1}\varphi)_{\Omega_R} = a_{s,\Omega_R}(\tilde{e}, v) &= a_{s,\Omega_R}(\tilde{e}, v - I_h v) \\
&\le c\sum_J \|\tilde{e}\|_{H^1(\Omega_J)}\|v - I_h v\|_{H^1(\Omega_J)}.
\end{aligned} \tag{4.27}$$

We distinguish the two cases $J \le I - 3$ and $J = I, I - 1, I - 2$ and begin with $J \le I - 3$. Notice, that $v \in H^2(\Omega_J')$ since no singularities occur at corners where Dirichlet and Neumann boundary intersect as long as the interior angle is smaller or equal to $\pi/2$, see [25]. It follows from standard interpolation theory

$$\|v - I_h v\|_{H^1(\Omega_J)}^2 \le \sum_{T \in \Omega_{J,h}} \|v - I_h v\|_{H^1(T)}^2 \le c\sum_{T \in \Omega_{J,h}} h_J^2 |v|_{H^2(T)}^2 \le ch_J^2 |v|_{H^2(\Omega_J')}^2. \tag{4.28}$$

To estimate the H^2-norm on the right-hand side we introduce a smooth cut-off function η_J with $\eta_J \equiv 1$ in Ω_J', $\eta_J \equiv 0$ in $\Omega\backslash\Omega_J''$ and $\|D^\alpha\eta_J\|_{L^\infty(\Omega_J'')} \le cd_J^{-|\alpha|}$. It follows

$$\begin{aligned}
\|v\|_{H^2(\Omega_J')} &\le \|\eta_J v\|_{H^2(\Omega)} \\
&\le c\| - \nabla \cdot A(x)\nabla(\eta_J v) + a_1(x)\nabla(\eta_J v) + a_0(x)\eta_J v\|_{L^2(\Omega_J'')} \\
&\le c\|\eta_J(-\nabla \cdot A(x)\nabla v + a_1(x)\nabla v + a_0(x)v)\|_{L^2(\Omega_J'')} \\
&\quad + c\|\nabla v \cdot A(x)\nabla\eta_J + v\nabla \cdot (A(x)\nabla\eta_J) + \nabla\eta_J A(x)\nabla v + a_1(x)v\nabla\eta_J\|_{L^2(\Omega_J'')} \\
&\le c\|(|x| + d_I)^{-1}\varphi\|_{L^2(\Omega_J'')} + cd_J^{-1}\|\nabla v\|_{L^2(\Omega_J'')} + cd_J^{-2}\|v\|_{L^2(\Omega_J'')}.
\end{aligned} \tag{4.29}$$

where the constants c depend on $\|A\|_{W^{1,\infty}(\Omega_R)}$, $\|a_1\|_{L^\infty(\Omega_R)}$ and $\|a_0\|_{L^\infty(\Omega_R)}$ but not on J. Since v admits Dirichlet boundary conditions on parts of $\partial\Omega_J''$ we can apply the Poincaré inequality and get

$$\|v\|_{L^2(\Omega_J'')} \leq cd_J\|\nabla v\|_{L^2(\Omega_J'')}.$$

Taking $\|\varphi\|_{L^2(\Omega_R)} = 1$ into account we can continue from estimate (4.29) with

$$\|v\|_{H^2(\Omega_J')} \leq cd_J^{-1} + cd_J^{-1}\|\nabla v\|_{L^2(\Omega_J'')}. \tag{4.30}$$

This yields together with inequality (4.28)

$$\|v - I_h v\|_{H^1(\Omega_J)} \leq ch_J d_J^{-1}\left(1 + \|\nabla v\|_{L^2(\Omega_J'')}\right).$$

With this estimate we apply Lemma 4.12 to the terms of the sum (4.27) and conclude for $J \leq I - 3$

$$
\begin{aligned}
\|\tilde{e}\|_{H^1(\Omega_J)}\|v - I_h v\|_{H^1(\Omega_J)} &\leq h_J d_J^{-1}\left(1 + \|\nabla v\|_{L^2(\Omega_J'')}\right)\|\tilde{e}\|_{H^1(\Omega_J)} \\
&\leq ch_J d_J^{-1}\left(1 + \|\nabla v\|_{L^2(\Omega_J'')}\right)\left(hd_J^{1-\beta-\mu}|y|_{V_\beta^{2,2}(\Omega_J'')}\right. \\
&\qquad \left. + h^{1+2\delta'/\lambda}d_J^{2-\gamma-\mu}|y|_{V_\gamma^{2,\infty}(\Omega_J'')} + \left\|d_J^{-1}\tilde{e}\right\|_{L^2(\Omega_J')}\right) \\
&\leq ch^2\left(1 + \|\nabla v\|_{L^2(\Omega_J'')}\right)\left(d_J^{1-\beta-2\mu}|y|_{V_\beta^{2,2}(\Omega_J'')}+\right. \\
&\qquad \left. h^{2\delta'/\lambda}d_J^{2-\gamma-2\mu}|y|_{V_\gamma^{2,\infty}(\Omega_J'')}\right) \\
&\qquad + ch_J d_J^{-1}\left(1 + \|\nabla v\|_{L^2(\Omega_J'')}\right)\left\|d_J^{-1}\tilde{e}\right\|_{L^2(\Omega_J')} \\
&\leq ch^2\left(1 + \|\nabla v\|_{L^2(\Omega_J'')}\right)\left(|y|_{V_\beta^{2,2}(\Omega_J'')} + h^{2\delta'/\lambda}|y|_{V_\gamma^{2,\infty}(\Omega_J'')}\right) \\
&\qquad + ch_J d_J^{-1}\left(1 + \|\nabla v\|_{L^2(\Omega_J'')}\right)\left\|d_J^{-1}\tilde{e}\right\|_{L^2(\Omega_J')},
\end{aligned}
\tag{4.31}
$$

since $1 - \beta - 2\mu = 2\delta' - \delta > 0$ and $2 - \gamma - 2\mu = \lambda - 2\mu = \delta' > 0$.

Let us now consider the case of $J = I, I-1, I-2$. We introduce the domains

$$\Omega_{J,h} = \{T : T \cap \Omega_J \neq \emptyset\}.$$

For $T \in \Omega_{J,h}$ one gets from [16, Proof of Theorem 3.2] the estimate

$$\|v - I_h v\|_{H^1(T)} \leq ch_J^{1-\beta}|v|_{V_\beta^{2,2}(T)}.$$

Summing up over all elements yields

$$\|v - I_h v\|_{H^1(\Omega_J)} \leq \|v - I_h v\|_{H^1(\Omega_{J,h})} \leq ch_J^{1-\beta}|v|_{V_\beta^{2,2}(\Omega_{J,h})} < ch_J^{1-\beta}|v|_{V_\beta^{2,2}(\Omega_J')}. \tag{4.32}$$

With a similiar argumentation as in the derivation of (4.29) and an a priori estimate for boundary value problems in domains with conical points (see [77]) it follows

$$
\begin{aligned}
|v|_{V^{2,2}_\beta(\Omega'_J)} &\le c \| (|x| + d_I)^{-1} \varphi \|_{V^{0,2}_\beta(\Omega''_J)} + c d_J^{-1} \| \nabla v \|_{V^{0,2}_\beta(\Omega''_J)} + c d_J^{-2} \| v \|_{V^{0,2}_\beta(\Omega''_J)} \\
&\le c d_J^{\beta-1} \| \varphi \|_{L^2(\Omega''_J)} + c d_J^{-1+\beta} \| \nabla v \|_{L^2(\Omega''_J)} + c d_J^{-2+\beta} \| v \|_{L^2(\Omega''_J)} \\
&\le c d_J^{\beta-1} + c d_J^{\beta-1} \| \nabla v \|_{L^2(\Omega''_J)}
\end{aligned}
$$

where we used the Poincaré inequality in the last step. This yields together with (4.32) and the same argumentation as above

$$
\| v - I_h v \|_{H^1(\Omega_J)} \le c \left(h_J d_J^{-1} \right)^{1-\beta} \left(1 + \| \nabla v \|_{L^2(\Omega''_J)} \right).
$$

We apply again Lemma 4.12 to the terms of the sum (4.27) and arrive at

$$
\begin{aligned}
\| \tilde{e} \|_{H^1(\Omega_J)} \| v - I_h v \|_{H^1(\Omega_J)} &\le c \left(h_J d_J^{-1} \right)^{1-\beta} \left(1 + \| \nabla v \|_{L^2(\Omega''_J)} \right) \| \tilde{e} \|_{H^1(\Omega_J)} \\
&\le c \Big[\left(h_J d_J^{-1} \right)^{1-\beta} \left(1 + \| \nabla v \|_{L^2(\Omega''_J)} \right) h^2 \| y \|_{V^{2,\infty}_\gamma(\Omega''_J)} \\
&\quad + \left(h_J d_J^{-1} \right)^{1-\beta} \left(1 + \| \nabla v \|_{L^2(\Omega''_J)} \right) \left\| d_J^{-1} \tilde{e} \right\|_{L^2(\Omega'_J)} \Big].
\end{aligned}
$$

In the following we use the estimate

$$
\| \nabla v \|_{L^2(\Omega)} \le c |\ln h| \tag{4.33}
$$

which is shown at the end of this proof. With this estimate and the application of Lemma 4.7 one has for $J = I, I-1, I-2$

$$
\| \tilde{e} \|_{H^1(\Omega_J)} \| v - I_h v \|_{H^1(\Omega_J)} \le c h^2 \| y \|_{V^{2,\infty}_\gamma(\Omega''_J)} + \left(h_J d_J^{-1} |\ln h|^{1/(1-\beta)} \right)^{1-\beta} \left\| d_J^{-1} \tilde{e} \right\|_{L^2(\Omega'_J)}. \tag{4.34}
$$

It follows from (4.26) and (4.27) with (4.31) and (4.34)

$$
\begin{aligned}
\| (|x| + d_I)^{-1} \tilde{e} \|_{L^2(\Omega_R)} &\le c \sum_{J=1}^{I} \| \tilde{e} \|_{H^1(\Omega_J)} \| v - I_h v \|_{H^1(\Omega_J)} \\
&\le c h^2 \sum_{J=1}^{I-3} \left(1 + \| \nabla v \|_{L^2(\Omega''_J)} \right) \left(|y|_{V^{2,2}_\beta(\Omega''_J)} + h^{2\delta'/\lambda} |y|_{V^{2,\infty}_\gamma(\Omega''_J)} \right) \\
&\quad + c \sum_{J=1}^{I-3} h_J d_J^{-1} \left(1 + \| \nabla v \|_{L^2(\Omega''_J)} \right) \left\| d_J^{-1} \tilde{e} \right\|_{L^2(\Omega'_J)} \\
&\quad + c \sum_{J=I-2}^{I} \left(h^2 \| y \|_{V^{2,\infty}_\gamma(\Omega''_J)} + \left(h_J d_J^{-1} |\ln h|^{1/(1-\beta)} \right)^{1-\beta} \left\| d_J^{-1} \tilde{e} \right\| \right).
\end{aligned}
$$

The application of the Cauchy-Schwarz inequality yields with $\left(\sum_{J=1}^{I-3}1\right)^{1/2}\sim|\ln h|^{1/2}$

$$\|(|x|+d_I)^{-1}\tilde{e}\|_{L^2(\Omega_R)}\leq ch^2\,|\ln h|^{1/2}\left[\left(\sum_{J=1}^{I-3}|y|^2_{V_\beta^{2,2}(\Omega''_J)}\right)^{1/2}+\left(\sum_{J=1}^{I-3}h^{4\delta'/\lambda}|y|^2_{V_\gamma^{2,\infty}(\Omega''_J)}\right)^{1/2}\right]$$

$$+\,ch^2\left(\sum_{J=1}^{I-3}\|\nabla v\|^2_{L^2(\Omega''_J)}\right)^{1/2}\cdot$$

$$\left[\left(\sum_{J=1}^{I-3}|y|^2_{V_\beta^{2,2}(\Omega''_J)}\right)^{1/2}+\left(\sum_{J=1}^{I-3}h^{4\delta'/\lambda}|y|^2_{V_\gamma^{2,\infty}(\Omega''_J)}\right)^{1/2}\right]$$

$$+\,c\,|\ln h|^{1/2}\left(\sum_{J=1}^{I-3}\left(h_Jd_J^{-1}\,|\ln h|\right)^2\|d_J^{-1}\tilde{e}\|^2_{L^2(\Omega'_J)}\right)^{1/2}+ch^2\|y\|^2_{V_\gamma^{2,\infty}(\Omega)}$$

$$+\left(\sum_{J=I-2}^{I}\left(h_Jd_J^{-1}\,|\ln h|^{1/(1-\beta)}\right)^{2(1-\beta)}\|d_J^{-1}\tilde{e}\|^2_{L^2(\Omega'_J)}\right)^{1/2}.$$

With Lemma 4.7 it follows for an arbitrary, but fixed $c_0<1$ and h small enough together with estimate (4.33)

$$\|(|x|+d_I)^{-1}\tilde{e}\|_{L^2(\Omega_R)}\leq ch^2\,|\ln h|\left(\|y\|_{V_\beta^{2,2}(\Omega)}+h^{2\delta'/\lambda}\,|\ln h|^{1/2}\,\|y\|_{V_\gamma^{2,\infty}(\Omega)}\right)$$
$$+\,ch^2\|y\|^2_{V_\gamma^{2,\infty}(\Omega)}+c_0\|(|x|+d_I)^{-1}\tilde{e}\|_{L^2(\Omega_R)}$$

Since $h^{2\delta'/\lambda}\,|\ln h|^{1/2}<c$ for h small one gets

$$\|(|x|+d_I)^{-1}\tilde{e}\|_{L^2(\Omega_R)}\leq ch^2\,|\ln h|\left(\|y\|_{V_\beta^{2,2}(\Omega)}+\|y\|_{V_\gamma^{2,\infty}(\Omega)}\right)+c_0\|(|x|+d_I)^{-1}\tilde{e}\|_{L^2(\Omega_R)}$$

and therefore

$$\|(|x|+d_I)^{-1}\tilde{e}\|_{L^2(\Omega_R)}\leq ch^2\,|\ln h|\left(\|y\|^2_{V_\beta^{2,2}(\Omega)}+\|y\|^2_{V_\gamma^{2,\infty}(\Omega)}\right)$$

what proves the assertion.

It remains to prove (4.33). We do this with an idea already used in [3]. We introduce the abbreviation $\sigma(r):=r+d_I$. It follows

$$\|v\|^2_{H^1(\Omega_R)}\leq ca_{\Omega_R}(v,v)=c(\sigma^{-1}\varphi,v)=c(\varphi,\sigma^{-1}v)$$
$$\leq c\|\varphi\|_{L^2(\Omega_R)}\|\sigma^{-1}v\|_{L^2(\Omega_R)}=c\|\sigma^{-1}v\|_{L^2(\Omega_R)}. \tag{4.35}$$

In the next step we show

$$\|\sigma^{-1}v\|_{L^2(\Omega_R)}\leq c\,|\ln h|\,\|v\|_{H^1(\Omega_R)} \tag{4.36}$$

which yiels together with (4.35) the estimate (4.33). We define

$$\psi(r) := \frac{d_I}{r + d_I} + \ln(r + d_I).$$

Then one has

$$\frac{\mathrm{d}}{\mathrm{d}r}\left(\psi(r) - \psi(0)\right) = \frac{r}{\sigma^2(r)}.$$

With this definition and partial integration one can conclude

$$\|\sigma^{-1}v\|_{L^2(\Omega_R)}^2 = \int_0^\omega \int_0^R \sigma^{-2}v^2 r \,\mathrm{d}r\,\mathrm{d}\varphi = \int_0^\omega \int_0^R v^2 \frac{\mathrm{d}}{\mathrm{d}r}[\psi(r) - \psi(0)]\,\mathrm{d}r\,\mathrm{d}\varphi$$

$$= \int_0^\omega \left[(\psi(r) - \psi(0))\,v^2\right]_0^R \,\mathrm{d}\varphi - \int_0^\omega \int_0^R (\psi(r) - \psi(0))\,2v\partial_r v \,\mathrm{d}r\,\mathrm{d}\varphi.$$
$$(4.37)$$

We estimate the last two terms separately. It follows

$$\int_0^\omega \left[(\psi(r) - \psi(0))\,v^2\right]_0^R \,\mathrm{d}\varphi = \int_0^\omega \left(\ln(R + d_I) - \ln d_I + \frac{d_I}{R + d_I} - 1\right) v^2(R,\varphi)\,\mathrm{d}\varphi$$

$$\leq \int_0^\omega (\ln(R + d_I) - \ln d_I)\,v^2(R,\varphi)\,\mathrm{d}\varphi$$

$$\leq (c + |\ln d_I|) \int_0^\omega v^2(R,\varphi)\,\mathrm{d}\varphi \leq (c + |\ln d_I|)\,\|v\|_{L^2(\partial\Omega_R)}^2.$$

With help of a trace theorem [34, Theorem 1.6.6] we can continue for h small enough

$$\int_0^\omega \left[(\psi(r) - \psi(0))\,v^2\right]_0^R \,\mathrm{d}\varphi \leq (c + |\ln d_I|)\,\|v\|_{L^2(\Omega_R)}\|v\|_{H^1(\Omega_R)}$$

$$\leq (c + |\ln d_I|)\,\|\sigma\|_{L^\infty(\Omega_R)}\|\sigma^{-1}v\|_{L^2(\Omega_R)}\|v\|_{H^1(\Omega_R)}$$

$$\leq c\,|\ln h|\,\|\sigma^{-1}v\|_{L^2(\Omega_R)}\|v\|_{H^1(\Omega_R)}.$$
$$(4.38)$$

where we have used $d_I \sim h^{2/\lambda}$ in the last step. In order to prove an estimate for the second term of the right-hand side of (4.37) we first prove the auxiliary result

$$\left|\frac{\psi(r) - \psi(0)}{r}\right| \leq \frac{c}{\sigma}\,|\ln h|.$$
$$(4.39)$$

We distinguish the cases $r \geq d_I$ and $r < d_I$ and begin with $r \geq d_I$. It is

$$|\psi(0)| = |1 + \ln d_I| \leq c\,|\ln d_I| = c\,|\ln h|,$$
$$(4.40)$$

$$|\psi(R)| = \left|\frac{d_I}{R + d_I} + \ln(R + d_I)\right| \leq c\,|\ln R|$$
$$(4.41)$$

and therefore

$$|\psi(r) - \psi(0)| \leq 2 \max_{0 \leq s \leq R} \psi(s) = c\,|\ln h|.$$

Since $r^{-1} \leq c\,\sigma^{-1}$ inequality (4.39) follows. For the case $r < d_I$ we use the mean value theorem and conclude

$$\left| \frac{\psi(r) - \psi(0)}{r} \right| \leq \max_{0 \leq s \leq d_I} |\psi'(s)| = \max_{0 \leq s \leq d_I} \left| \frac{s}{\sigma(s)^2} \right|.$$

As the last function is monotonically increasing in $[0, d_I]$ one can estimate

$$\left| \frac{\psi(r) - \psi(0)}{r} \right| \leq \left| \frac{d_I}{\sigma(d_I)^2} \right| = \frac{1}{4 d_I} \leq \frac{c}{\sigma}$$

and inequality (4.39) follows. With the help of this estimate one can conclude for the second term of the right-hand side in equation (4.37)

$$\int_0^\omega \int_0^R (\psi(r) - \psi(0))\, 2v \partial_r v \, \mathrm{d}r \, \mathrm{d}\varphi \leq c \, |\ln h| \int_0^\omega \int_0^R \sigma^{-1} r v \partial_r v \, \mathrm{d}r \, \mathrm{d}\varphi$$
$$\leq c \, |\ln h| \, \|\sigma^{-1} v\|_{L^2(\Omega_R)} \|v\|_{H^1(\Omega_R)}.$$

This yields together with (4.38) and (4.37) the estimate (4.36) and the assertion (4.33) follows. $\qquad\square$

Remark 4.14. If one denotes by $h_{J'}$ the element size in Ω'_J and by $h_{J''}$ the element size in Ω''_J one has

$$h_J \sim \frac{1}{2} h_{J'} \sim \frac{1}{4} h_{J''}.$$

Therefore in Lemmata 4.12 and 4.13 one can substitute Ω_J by Ω'_J and Ω'_J by Ω''_J.

Now we are able to prove Theorem 4.4.

Proof of Theorem 4.4. First we assume that $y - y_h$ admits its maximum at a point x_0 which is contained in Ω_R. For all other domains Ω_{R_i} the argumentation is the same. We can estimate

$$\|y - y_h\|_{L^\infty(\Omega_R)} = \|\tilde{y} - y_h\|_{L^\infty(\Omega_R)}$$
$$\leq \|\tilde{y} - \tilde{y}_h\|_{L^\infty(\Omega_R)} + \|\tilde{y}_h - y_h\|_{L^\infty(\Omega_R)}. \qquad (4.42)$$

Since

$$a(\tilde{y}_h - y_h, \chi) = 0 \quad \forall \chi \in V_h|_{\tilde{\Omega}_R}$$

we can write for the second term of the right-hand side of the last inequality with Theorem 3.1 of [117] and the triangle inequality

$$\|\tilde{y}_h - y_h\|_{L^\infty(\Omega_R)} \leq c \, |\ln h|^{1/2} \, \|\tilde{y}_h - y_h\|_{L^2(\tilde{\Omega}_R)}$$
$$\leq c \, |\ln h|^{1/2} \left(\|\tilde{y} - \tilde{y}_h\|_{L^2(\tilde{\Omega}_R)} + \|y - y_h\|_{L^2(\tilde{\Omega}_R)} \right).$$

In Theorem 3.1 of [117] there is the factor $h^{-\varepsilon}$ written instead of $|\ln h|^{1/2}$. At the beginning of Section 3 of that paper it is mentioned that one can replace $h^{-\varepsilon}$ by $|\ln h|^{1/2}$ as long as one can prove the Sobolev inequality with this factor as we did in Lemma 4.9, compare also the corresponding proof in [117, p. 95]. It follows from Theorem 4.3

$$\|\tilde{y} - \tilde{y}_h\|_{L^2(\tilde{\Omega}_R)} + \|y - y_h\|_{L^2(\tilde{\Omega}_R)} \le ch^2 \|f\|_{L^2(\Omega)}$$

such that

$$\|\tilde{y}_h - y_h\|_{L^\infty(\Omega_R)} \le ch^2 |\ln h|^{1/2} \|f\|_{L^2(\Omega)}.$$

It remains estimate the first term of (4.42). From Lemma 4.10 we have together with Lemmas 4.12 and 4.13 for $J = I, I-1, I-2$ the estimate

$$\|y - y_h\|_{L^\infty(\Omega_J)} \le ch^2 |\ln h|^{3/2} \left(\|y\|_{V_\beta^{2,2}(\Omega)} + \|y\|_{V_\gamma^{2,\infty}(\Omega)} \right).$$

For Ω_J, $J \ne I, I-1, I-2$ we conclude from Lemmas 4.10 and 4.13 the estimate

$$\|y - y_h\|_{L^\infty(\Omega_J)} \le ch^2 |\ln h| \left(\|y\|_{V_\beta^{2,2}(\Omega)} + \|y\|_{V_\gamma^{2,\infty}(\Omega)} \right).$$

The last two inequalities yield together with the a priori estimates of Lemmata 4.1 and 4.2 and the embedding $C^{0,\sigma}(\bar{\Omega}) \hookrightarrow L^2(\Omega)$.

$$\|y - y_h\|_{L^\infty(\Omega_R)} \le ch^2 |\ln h|^{3/2} \|f\|_{C^{0,\sigma}(\bar{\Omega})}.$$

For the case $x_0 \in \Omega_0$ we can conclude from [116, Theorem 5.1]

$$\|y - y_h\|_{L^\infty(\Omega_0)} \le c |\ln h| \|y - I_h y\|_{L^\infty(\Omega)} + \|y - y_h\|_{L^2(\Omega)}$$
$$\le ch^2 |\ln h| \|f\|_{C^{0,\sigma}(\bar{\Omega})}$$

where we have used an interpolation estimate

$$\|y - I_h y\|_{L^\infty(\Omega)} \le ch^2 \|y\|_{V_\gamma^{2,\infty}(\Omega)}$$

as it is shown in the proof of Lemma 4.10, the a priori estimate of Lemma 4.2, Theorem 4.3 and the embedding $C^{0,\sigma}(\bar{\Omega}) \hookrightarrow L^2(\Omega)$. $\qquad\square$

4.2 Scalar elliptic equations in prismatic domains

In this section we consider the boundary value problem

$$Ly = f \quad \text{in } \Omega, \qquad By = 0 \quad \text{on } \Gamma = \partial\Omega. \tag{4.43}$$

We analyze two different cases, namely pure Dirichlet boundary conditions, i.e.,

$$L = -\Delta, \quad B = \text{Id}, \tag{4.44}$$

with variational formulation

$$\text{Find } y \in V_0: \qquad a_D(y, v) = (f, v)_{L^2(\Omega)} \qquad \forall v \in V_0 \tag{4.45}$$

where the bilinear form $a_D : H^1(\Omega) \times H^1(\Omega) \to \mathbb{R}$ is defined as

$$a_D(y, v) = \int_\Omega \nabla y \cdot \nabla v,$$

and pure Neumann boundary conditions, i.e.,

$$L = -\Delta + \mathrm{Id}, \quad B = \frac{\partial}{\partial n}. \tag{4.46}$$

with variational formulation

$$\text{Find } y \in H^1(\Omega): \quad a_N(y, v) = (f, v)_{L^2(\Omega)} \quad \forall v \in H^1(\Omega) \tag{4.47}$$

where the bilinear form $a_N : H^1(\Omega) \times H^1(\Omega) \to \mathbb{R}$ is defined as

$$a_N(y, v) = \int_\Omega \nabla y \cdot \nabla v + \int_\Omega y \cdot v.$$

Here, $\Omega = G \times Z \subset \mathbb{R}^3$ is a domain with boundary $\partial\Omega$, where $G \subset R^2$ is a bounded polygonal domain and $Z := (0, z_0) \subset \mathbb{R}$ is an interval. It is assumed that the cross-section G has only one corner with interior angle $\omega > \pi$ at the origin; thus Ω has only one "singular edge" which is part of the x_3-axis. Situation with more than one critical edge can be reduced to this case by a localization argument, see, e.g., [80].

The results of this section are published in [17] and [18].

4.2.1 Regularity

There are many publications about the regularity of solutions of elliptic boundary value problems in domains with edges, especially for the Dirichlet case. Let us first concentrate on this case. The crucial idea of the investigation of the Dirichlet problem in a prismatic domain is to reduce it by a real Fourier transformation to a plane problem. Indeed, one can show that the regularity of the solution y of (4.45) is determined by a plane Dirichlet problem in a cone, which is of the same type as the one described in Section 4.1, [80, III.7]. This allows to formulate the following regularity result. Notice, that the result is actually the same as in Lemma 4.1 for the two-dimensional setting. We only have substituted the eigenvalue λ of by π/ω since we consider only the Laplace operator.

Lemma 4.15. *Let p and β be given real numbers with $p \in (1, \infty)$ and $\beta > 2 - \pi/\omega - 2/p$. Moreover, let f be a function in $V_\beta^{0,p}(\Omega)$. Then the solution of the boundary value problem (4.45) belongs to $H_0^1(\Omega) \cap V_\beta^{2,p}(\Omega)$. Moreover, the inequality*

$$\|y\|_{V_\beta^{2,p}(\Omega)} \le c\|f\|_{V_\beta^{0,p}(\Omega)}$$

is valid.

Proof. With $\mathrm{Im}\lambda_- -- -\pi/\omega$ the assertion follows from Lemmata 4 and 5 of [115]. $\qquad\square$

The drawback of describing the solution in the $V_\beta^{k,p}(\Omega)$-spaces is, that the space $H^1(\Omega)$ does not belong to the scale of these weighted Sobolev spaces. Particularly, a solution of the Neumann problem may not be included in $H^1(\Omega) \cap V_0^{1,2}(\Omega)$, think e.g. of the function $u \equiv 1$. This is the reason why problem (4.47) is not included in the paper [16], where the authors demand $u \in H^1(\Omega) \cap V_0^{1,2}(\Omega)$. A way out is the description of the solution of (4.47) in the spaces $W_\beta^{k,p}(\Omega)$.

In contrast to the Dirichlet case there are not too many publications on regularity results concerning Neumann problems in domains with edges. Therefore we give a short overview of the literature concerning this topic and start with the book of Grisvard [63], where estimates on the solution of the Neumann problem for the Laplace equation and the Lamé system in Sobolev and Sobolev-Slobodeckiĭ spaces with $p = 2$ and without weight are given. Dauge [47] proved regularity results for linear elliptic Neumann problems in L^p Sobolev spaces without weight. Maz'ya and Roßmann obtained regularity results in weighted Sobolev spaces in a cone for general p. Their result about the Neumann problem in a dihedron requires additional regularity on the solution, which cannot be guaranteed in our case. Zaionchkovskii and Solonnikov [130], Roßmann [114] and Nazarov and Plamenevskiĭ [99] proved solvability theorems and regularity results for the Neumann problem in weighted Sobolev spaces for $p = 2$. With the results of Zaionchkovskii and Solonnikov [130] we obtain the following theorem.

Lemma 4.16. *Let u be the solution of* (4.47). *If $f \in W_\beta^{0,2}(\Omega)$ with $\beta > 1 - \pi/\omega$, then u is contained in the space $W_\beta^{2,2}(\Omega)$ and satisfies the inequality*

$$\|y\|_{W_\beta^{2,2}(\Omega)} \le c\|f\|_{W_\beta^{0,2}(\Omega)}.$$

Proof. We first consider problem (4.47) in a dihedron $\mathcal{D}_\omega = \{x = (x', x_3) : x' \in K, x_3 \in \mathbb{R}\}$ where K denotes an angle of the form $\{x' = (x_1, x_2) \in \mathbb{R}^2 : 0 < r < \infty, \ 0 < \varphi < \omega\}$ in polar coordinates r, φ. Setting $k = 0$ in Theorem 5.2 of [130] we can conclude

$$|y|_{W_\beta^{2,2}(\mathcal{D}_\omega)} + \|y\|_{W^{1,2}(\mathcal{D}_\omega)} \le c\|f\|_{W_\beta^{0,2}(\mathcal{D}_\omega)}. \tag{4.48}$$

Since $\beta > 0$ estimate (4.48) keeps valid if one substitutes the left-hand side of the inequality by $\|y\|_{W_\beta^{2,2}(\mathcal{D}_\omega)}$. Problem (4.47) can be locally transformed near an edge point by a diffeomorphism into a boundary value problem in the dihedron $\mathcal{D}_\omega$. By the use of a partion of unity method one can fit together the local results to obtain the result for the domain Ω. Details on this technique can be found e.g. in the book of Kufner and Sändig [80, Section 8]. □

Up to know it is not reflected in the regularity results that there is some "additional smoothness" along the edge for the solution of both, the Dirichlet and the Neumann problem. In order to exploit this feature, we follow the explanations of Grisvard in [64] and split the solution of (4.45) and (4.47), respectively, in a singular part y_s and a regular part y_r. In detail, this means that the solution y can be written for $f \in L^p(\Omega)$, $2 \le p < \infty$ as

$$y = y_s + y_r, \tag{4.49}$$

where $y_r \in W^{2,p}(\Omega)$ and

$$y_s = \xi(r)\gamma(r, x_3)r^\lambda\Theta(\varphi) \quad \text{with } \lambda = \frac{\pi}{\omega}.$$

Here $\xi(r)$ is a smooth cut-off function and $\Theta(\varphi) = \sin \lambda\varphi$ for the Dirichlet boundary conditions and $\Theta(\varphi) = \cos \lambda\varphi$ for the Neumann boundary conditions. The coefficient function γ can be written as a convolution integral,

$$\gamma(r, x_3) = \frac{1}{\pi}\int_{\mathbb{R}} \frac{r}{r^2 + s^2}q(x_3 - s)\,ds$$

where the smoothness of the function q can be characterized in Besov spaces depending on λ. This decomposition allows to formulate the following lemma.

Lemma 4.17. *Let y be the solution of (4.45) or (4.47) for a right-hand side $f \in L^p(\Omega)$, $2 \le p < \infty$. For the singular part y_s the inequalities*

$$\|r^\beta \partial_{ij}y_s\|_{L^p(\Omega)} + \|\partial_{3i}y_s\|_{L^p(\Omega)} + \|\partial_{33}y_s\|_{L^p(\Omega)} \le c\|f\|_{L^p(\Omega)}, \quad i,j = 1,2 \tag{4.50}$$

$$\|r^{\beta-1}\partial_i y_s\|_{L^p(\Omega)} + \|r^{-1}\partial_3 y_s\|_{L^p(\Omega)} \le c\|f\|_{L^p(\Omega)}, \quad i = 1,2 \tag{4.51}$$

$$\|r^{\beta-2}y_s\|_{L^p(\Omega)} \le c\|f\|_{L^p(\Omega)} \tag{4.52}$$

are valid for

$$\beta > 2 - \frac{2}{p} - \lambda \quad \text{if } 1 - \frac{2}{p} < \lambda \le 2 - \frac{2}{\mu} \quad \text{and}$$

$$\beta = 0 \quad \text{if } \lambda > 2 - \frac{2}{p}.$$

For the regular part y_r the estimate

$$\|y_r\|_{W^{2,p}(\Omega)} \le c\|f\|_{L^p(\Omega)} \tag{4.53}$$

holds.

Proof. In [9, Section 2.1] the assertions (4.50)–(4.52) are proved for the Dirichlet problem. In order to get the estimates for the Neumann problem one just has to replace $\sin\left(\frac{j\pi\varphi}{\omega}\right)$ by $\cos\left(\frac{j\pi\varphi}{\omega}\right)$ in that proof. Expression (4.53) follows from [64, Theorem 6.6]. □

Corollary 4.18. *Let y be the solution of (4.45) and (4.47) respectively. Then one has*

$$\frac{\partial y}{\partial x_3} \in H^1(\Omega) \quad \text{and} \quad \left\|\frac{\partial y}{\partial x_3}\right\|_{H^1(\Omega)} \le c\|f\|_{L^2(\Omega)}.$$

Proof. Since $y \in H^1(\Omega)$ and $u_r \in H^2(\Omega)$ with $\|y_r\|_{H^2(\Omega)} \le c\|f\|_{L^2(\Omega)}$ the assertion follows from (4.49) and (4.50). □

55

Remark 4.19. For the Dirichlet problem the inequalities (4.50)–(4.52) are also valid for the regular part y_r (see [80]). This is not the case for the Neumann problem since the regular part needs not to vanish at the edge.

The solution y of the boundary value problem (4.45) or (4.47) is not contained in the space $W^{1,\infty}(\Omega)$. Instead, one has $r^\beta \nabla y \in L^\infty(\Omega)$ with a suitable weight β. A reasonable attempt to determine an appropriate value for the weight β is the use of Sobolev embedding theorems and Lemma 4.17. Let us quickly recapitulate the argumentation from [127] for the Dirichlet case.

For a right-hand side $f \in L^p(\Omega)$, $p > 3$ one has according to (4.51), (4.52) and Remark 4.19

$$r^\beta \nabla y \in W^{1,p}(\Omega) \quad \text{for } \beta > 2 - \frac{2}{p} - \lambda.$$

The embedding $W^{1,p}(\Omega) \hookrightarrow L^\infty(\Omega)$ yields

$$r^\beta \nabla y \in L^\infty(\Omega) \quad \text{for } \beta > 2 - \frac{2}{3} - \lambda = \frac{4}{3} - \lambda.$$

It turns out that these estimates based on embeddings theorems and regularity results for finite p are not sharp enough. The informal consideration $y \sim r^\lambda$ and consequently $\nabla y \sim r^{\lambda-1}$ suggests, that one can expect, that a weight $\beta > 1 - \lambda$ is already large enough. In fact, the following lemma shows, that this is true for both, Neumann and Dirichlet boundary conditions. To this end a more involved proof is necessary that uses regularity results in weighted Hölder spaces given by Maz'ya and Rossmann in [91].

Lemma 4.20. *Let y be the solution of (4.45) or (4.47) with a right-hand side $f \in C^{0,\sigma}(\bar{\Omega})$, $\sigma \in (0,1)$. Then the estimates*

$$\|r^\beta \nabla y\|_{L^\infty(\Omega)} \le c\|f\|_{C^{0,\sigma}(\bar{\Omega})}, \quad \beta > 1 - \lambda \tag{4.54}$$

$$\|\partial_3 y\|_{L^\infty(\Omega)} \le c\|f\|_{C^{0,\sigma}(\bar{\Omega})} \tag{4.55}$$

hold true.

Proof. In order to prove the assertion (4.54), we use the results from [91, Subsection 5.3]. From Theorem 5.1 and its proof in that paper, one has the a priori estimate

$$\|y\|_{C^{2,\sigma}_{\gamma,\delta}(\Omega)} \le c\|f\|_{C^{0,\sigma}_{\gamma,\delta}(\Omega)}. \tag{4.56}$$

In the case of our prismatic domain the norm in $C^{l,\sigma}_{\gamma,\delta}(\Omega)$ that is given in [91] reduces to

$$\|y\|_{C^{l,\sigma}_{\gamma,\delta}(\Omega)} = \sum_{|\alpha| \le l} \sup_{x \in \Omega} (\rho_1(x)\rho_2(x))^{\gamma - l - \sigma + |\alpha|} \left(\frac{r(x)}{\rho(x)}\right)^{H(\delta - l - \sigma + |\alpha|)} |\partial^\alpha y(x)|$$

$$+ \sum_{k=1}^{2} \sum_{|\alpha| = l - k_1} \sup_{x_1, x_2 \in \Omega} \rho_k(x_1)^{\gamma - \delta} \frac{|\partial^\alpha y(x_1) - \partial^\alpha y(x_2)|}{|x_1 - x_2|^{k_1 + \sigma - \delta}} \tag{4.57}$$

$$+ \sum_{|\alpha| = l} \sup_{|x_1 - x_2| < r(x_1)/2} \rho_1(x_1)^\gamma \rho_2(x_1)^\gamma \left(\frac{r(x_1)}{\rho(x_1)}\right)^\delta \frac{|\partial^\alpha y(x_1) - \partial^\alpha y(x_2)|}{|x_1 - x_2|^\sigma}.$$

The second term only appears in case of Neumann boundary conditions. Here, l is nonnegative integer that serves as differentiability exponent, while $0 < \sigma < 1$ is a Hölder exponent. The weights γ and δ correspond to corner and edge singularities. The functions $\rho_1(x)$ and $\rho_2(x)$ denote the distance of x to the corners, $r(x)$ is the distance of x to the edge and $\rho(x) = \min(\rho_1(x), \rho_2(x))$. Further, $k_1 = [\delta - \sigma] + 1$, where $[\delta - \sigma]$ denotes the greatest integer less or equal to $\delta - \sigma$. The function H is defined as $H(t) = t$ for Dirichlet boundary conditions and as $H(t) = \max(t, 0)$ for Neumann boundary conditions. In the prismatic domain no corner singularities occur such that we can choose $\gamma = \delta$ with the conditions

$$2 - \lambda + \sigma < \gamma < 2 + \sigma \qquad (4.58)$$

and $\gamma - \sigma \neq 1$ (comp. [91, prior to Theorem 5.1]). Now we can reduce our considerations concerning the norm in $C^{2,\sigma}_{\gamma,\delta}(\Omega)$ on the first term and $|\alpha| = 1$. Taking $\gamma = \delta$ into account, the relevant part is

$$M := \sum_{|\alpha|=1} \sup_{x \in \Omega} (\rho_1(x)\rho_2(x))^{\gamma-1-\sigma} \left(\frac{r(x)}{\rho(x)} \right)^{H(\gamma-1-\sigma)} |\partial^\alpha y(x)|.$$

Using inequality (4.58) it follows

$$\gamma - 1 - \sigma > 2 - \lambda - 1 = 1 - \lambda > 0 \qquad (4.59)$$

since $\lambda \in \left(\frac{1}{2}, 1 \right)$. Therefore $H(\gamma - 1 - \sigma) = \gamma - 1 - \sigma$ in both cases, Dirichlet and Neumann boundary condition. Now we introduce the domains $\Omega_1 = \{x \in \Omega, \rho(x) = \rho_1(x)\}$ and $\Omega_2 = \{x \in \Omega, \rho(x) = \rho_2(x)\}$. For every α with $|\alpha| = 1$, one can write

$$\sup_{x \in \Omega} (\rho_1(x)\rho_2(x))^{\gamma-1-\sigma} \left(\frac{r(x)}{\rho(x)} \right)^{\gamma-1-\sigma} |\partial^\alpha y(x)|$$

$$\geq \sup_{x \in \Omega_1} (\rho_1(x)\rho_2(x))^{\gamma-1-\sigma} \left(\frac{r(x)}{\rho(x)} \right)^{\gamma-1-\sigma} |\partial^\alpha y(x)|$$

$$= \sup_{x \in \Omega_1} \rho_2(x)^{\gamma-\sigma-1} r(x)^{\gamma-\sigma-1} |\partial^\alpha y(x)|$$

$$\geq c \cdot \sup_{x \in \Omega_1} r(x)^{\gamma-\sigma-1} |\partial^\alpha y(x)| \qquad (4.60)$$

since $\rho_2(x) \geq \frac{z_0}{2}$ for $x \in \Omega_1$. Analogously one has

$$\sup_{x \in \Omega} (\rho_1(x)\rho_2(x))^{\gamma-1-\sigma} \left(\frac{r(x)}{\rho(x)} \right)^{\gamma-1-\sigma} |\partial^\alpha y(x)| \geq c \cdot \sup_{x \in \Omega_2} r(x)^{\gamma-\sigma-1} |\partial^\alpha y(x)|. \qquad (4.61)$$

The estimates (4.60) and (4.61) yield

$$M \geq \|r^{\gamma-\sigma-1} \nabla y\|_{L^\infty(\Omega)}.$$

This entails for $\beta := \gamma - \sigma - 1$

$$\|r^\beta \nabla y\|_{L^\infty(\Omega)} \leq c\|y\|_{C^{2,\sigma}_{\gamma,\gamma}(\Omega)} \leq c\|f\|_{C^{0,\sigma}_{\gamma,\gamma}(\Omega)}, \quad \beta > 1 - \lambda, \qquad (4.62)$$

where we have used (4.56) and (4.59). In the following lines, we show

$$C^{0,\sigma}(\bar{\Omega}) \hookrightarrow C^{0,\sigma}_{\gamma,\gamma}(\Omega) \text{ for } \gamma - \sigma \geq 0. \tag{4.63}$$

The first term in the norm definition (4.57) yields for $l = 0$

$$\sup_{x\in\Omega} \rho_1(x)^{\gamma-\sigma} \rho_2(x)^{\gamma-\sigma} \left(\frac{r(x)}{\rho(x)}\right)^{H(\gamma-\sigma)} |y(x)| \leq c \cdot \sup_{x\in\Omega} r(x)^{\gamma-\sigma} |y(x)|$$

with the same argumentation as above. The second term vanishes since $l - k_1 < 0$. The third term results in

$$\sup_{|x_1-x_2|<r(x_1)/2} \rho_1(x)^{\gamma} \rho_2(x)^{\gamma} \left(\frac{r(x)}{\rho(x)}\right)^{\gamma} \frac{|y(x_1) - y(x_2)|}{|x_1 - x_2|^{\sigma}} \leq$$

$$c \cdot \sup_{|x_1-x_2|<r(x_1)/2} r(x_1)^{\gamma} \frac{|y(x_1) - y(x_2)|}{|x_1 - x_2|^{\sigma}}.$$

With $\gamma > \gamma - \sigma > 0$ these two estimates yield (4.63). Therefore the assertion (4.54) follows from (4.62). According to Lemma 4.17, one has $\partial_3 y \in W^{1,p}(\Omega)$. For $p > 3$ the Sobolev embedding $W^{1,p}(\Omega) \hookrightarrow L^{\infty}(\Omega)$ is valid. Therefore we can conclude

$$\|\partial_3 y\|_{L^{\infty}(\Omega)} \leq c \|\partial_3 y\|_{W^{1,p}(\Omega)} \leq c \|f\|_{L^p(\Omega)} \leq c \|f\|_{C^{0,\sigma}(\bar{\Omega})}$$

what is exactly assertion (4.55). $\qquad\square$

4.2.2 Finite element error estimates

We recall the definition of the discrete spaces V_h and V_{0h},

$$V_{0h} := \{v_h \in C(\bar{\Omega}) : v_h|_T \in \mathcal{P}_1 \text{ for all } T \in \mathcal{T}_h \text{ and } v_h = 0 \text{ on } \partial\Omega\},$$
$$V_h := \{v_h \in C(\bar{\Omega}) : v_h|_T \in \mathcal{P}_1 \text{ for all } T \in \mathcal{T}_h\},$$

for admissible triangulations $\mathcal{T}_h$, see page 16. The finite element approximation y_h of the solution of (4.45) is given as unique solution of the problem

$$\text{Find } y_h \in V_{0h} : \qquad a_D(y_h, v_h) = (f, v_h)_{L^2(\Omega)} \qquad \forall v_h \in V_{0h}. \tag{4.64}$$

Analogously, we have for the Neumann problem (4.47)

$$\text{Find } y_h \in V_h : \qquad a_N(y_h, v_h) = (f, v_h)_{L^2(\Omega)} \qquad \forall v_h \in V_h. \tag{4.65}$$

Notice, that in both cases the Lax-Milgram lemma guarantees existence and uniqueness of y_h.

The following two theorems provide estimates of the global interpolation error for the solutions of the boundary value problems (4.45) and (4.47) on anisotropic meshes.

Theorem 4.21. *Let y be the solution of (4.45) and E_{0h} the interpolation operator defined in (3.5). Then the estimate*

$$|y - E_{0h}y|_{H^1(\Omega)} \le ch\|f\|_{L^2(\Omega)} \tag{4.66}$$

holds if the mesh satisfies (2.13) with $\mu < \pi/\omega$.

Proof. The theorem can be proved along the lines of the proof of Theorem 14 of [6]. The necessary prerequisites are provided here with Lemmata 4.15, 4.17, Remark 4.19 and estimates (3.22) and (3.23) for $p = q = 2$. For the sake of completeness we sketch the details here. For estimating the error we distinguish between elements near the edge M and those away from M. Let us start with the elements T with $\overline{T} \cap M = \emptyset$. For those one knows that $y \in H^2(T)$ and thus we can use estimate (3.22) for $p = 2$, i.e.,

$$|y - E_{0h}y|_{H^1(T)} \le c \sum_{|\alpha|=1} h_T^\alpha |D^\alpha u|_{H^1(S_T)}. \tag{4.67}$$

Since

$$r_T \le \text{dist}(S_T, M) + h_{1,T} \sim \text{dist}(S_T, M) + h[\text{dist}(S_T, M)]^{1-\mu}$$

it follows for h sufficiently small

$$r_T \le c \cdot \text{dist}(S_T, M).$$

With this we can continue from (4.67) and get

$$|y - E_{0h}y|_{H^1(T)} \le c \left(\sum_{i=1}^{2} h_{i,T} r_T^{-\beta} \left|\frac{\partial y}{\partial x_i}\right|_{V_\beta^{1,2}(S_T)} + h_{3,T} \left|\frac{\partial y}{\partial x_3}\right|_{V_0^{1,2}(S_T)} \right) \tag{4.68}$$

for any $\beta > 1 - \pi/\omega$. Since $\mu < \pi/\omega$, the choice $\beta = 1 - \mu$ is admissible and we obtain for $r_T > 0$ from (2.13) the relation

$$h_{i,T} r_T^{-\beta} \sim h r_T^{1-\mu-\beta} = h \quad (i = 1, 2).$$

This yields together with (4.68)

$$|y - E_{0h}y|_{H^1(T)} \le ch \left(\sum_{i=1}^{2} \left|\frac{\partial y}{\partial x_i}\right|_{V_\beta^{1,2}(S_T)} + \left|\frac{\partial y}{\partial x_3}\right|_{V_0^{1,2}(S_T)} \right). \tag{4.69}$$

Consider now the elements T with $\overline{T} \cap M \ne \emptyset$. With the triangle inequality and the stability estimate (3.23) with $p = 2$ one gets for $\beta \in (1 - \pi/\omega, 1)$

$$|y - E_{0h}y|_{H^1(T)} \le |y|_{H^1(T)} + |E_{0h}y|_{H^1(T)}$$

$$\le c \left(\sum_{|\alpha|=1} \|D^\alpha y\|_{L^2(T)} + h_{1,T}^{-\beta} \sum_{|\alpha|-1} h_T^\alpha \|D^\alpha y\|_{V_\beta^{1,2}(S_T)} \right). \tag{4.70}$$

Taking into account that $r \leq c h_{1,T}$ in T and $1 - \beta > 0$ one obtains

$$\sum_{|\alpha|=1} \|D^\alpha y\|_{L^2(T)} \leq c \left(\sum_{i=1}^{2} h_{1,T}^{1-\beta} \left\| \frac{\partial y}{\partial x_i} \right\|_{V_{\beta-1}^{0,2}(T)} + h_{3,T} \left\| \frac{\partial y}{\partial x_3} \right\|_{V_{-1}^{0,2}(T)} \right)$$

$$\leq ch \left(\sum_{i=1}^{2} \left\| \frac{\partial y}{\partial x_i} \right\|_{V_\beta^{1,2}(T)} + \left\| \frac{\partial y}{\partial x_3} \right\|_{V_0^{1,2}(T)} \right). \tag{4.71}$$

In the last step we have used that $h_{1,T}^{1-\beta} \sim h^{(1-\beta)/\mu} = h$ for $\beta = 1 - \mu$ and $h_{3,T} \sim h$ (comp. (2.13)). For the second term in (4.70) we get with $r^\beta \leq h_{1,T}^\beta$

$$h_{1,T}^{-\beta} \sum_{|\alpha|=1} h_T^\alpha \|D^\alpha y\|_{V_\beta^{1,2}(S_T)} \leq c \left(\sum_{i=1}^{2} h_{1,T}^{1-\beta} \left\| \frac{\partial y}{\partial x_i} \right\|_{V_\beta^{1,2}(S_T)} + h_{1,T}^{-\beta} h \left\| \frac{\partial y}{\partial x_3} \right\|_{V_\beta^{1,2}(S_T)} \right)$$

$$\leq ch \left(\sum_{i=1}^{2} \left\| \frac{\partial y}{\partial x_i} \right\|_{V_\beta^{1,2}(S_T)} + \left\| \frac{\partial y}{\partial x_3} \right\|_{V_0^{1,2}(S_T)} \right). \tag{4.72}$$

Inserting (4.71) and (4.72) in (4.70) one can see that (4.69) is also valid for elements T with $\overline{T} \cap M \neq \emptyset$ with full norms instead of seminorms at the right-hand side. Summing up over all elements one gets

$$|y - E_{0h} y|_{H^1(\Omega)} \leq ch \left(\sum_{i=1}^{2} \left\| \frac{\partial y}{\partial x_i} \right\|_{V_\beta^{1,2}(\Omega)} + \left\| \frac{\partial y}{\partial x_3} \right\|_{V_0^{1,2}(\Omega)} \right)$$

with $\beta = 1 - \mu \in (1 - \pi/\omega, 1)$. Notice that this estimate is possible since only a finite number (independent of h) of patches S_T overlap. The application of Lemma 4.15 and Lemma 4.17 for $p = 2$ proves in view of Remark 4.19 estimate (4.66). $\qquad \square$

In case of Neumann boundary conditions we cannot prove a global estimate of the interpolation error in the same way as in Theorem 14 of [6] as we did it in the proof of Theorem 4.21. The reason is, that the solution y admits a different regularity in this case (comp. Lemma 4.15 and Lemma 4.16) and in particular does not vanish at the edge. Instead, the results of Theorem 3.7 play a keyrole.

Theorem 4.22. *Let y be the solution of (4.47) and E_h the interpolation operator defined in (3.2). Then the estimate*

$$\|y - E_h y\|_{H^1(\Omega)} \leq ch\|f\|_{L^2(\Omega)} \tag{4.73}$$

holds if the mesh satisfies (2.13) with $\mu < \pi/\omega$.

Proof. We use the estimates of the local error to get an estimate for the global error. Therefore we distinguish between elements next to the edge M and elements away from

M. We begin with the elements T with $\overline{T} \cap M = \emptyset$. Then $y \in H^2(T)$ and from (3.7) it follows with $p = 2$

$$\|y - E_h y\|_{H^1(T)} \leq c \sum_{|\alpha|=1} h_T^\alpha \|D^\alpha y\|_{H^1(S_T)}$$

$$\leq c \left(\sum_{i=1}^2 h_{i,T} r_T^{-\beta} \left\| \frac{\partial y}{\partial x_i} \right\|_{W_\beta^{1,2}(S_T)} + h_{3,T} \left\| \frac{\partial y}{\partial x_3} \right\|_{H^1(S_T)} \right)$$

for all $\beta < 1 - \pi/\omega$. For the last estimate we have used Lemma 4.17 and $r_T \leq ch_{3,T}$. Since $\mu < \pi/\omega$, the choice $\beta = 1 - \mu$ is admissible and we obtain from (2.13) the relation

$$h_{i,T} r_T^{-\beta} \sim h r_T^{1-\mu-\beta} = h \quad (i = 1, 2).$$

Combining this last estimates with the fact that $h_{3,T} \sim h$, one arrives at

$$\|y - E_h y\|_{H^1(T)} \leq c \left(h \sum_{i=1}^2 \left\| \frac{\partial y}{\partial x_i} \right\|_{W_\beta^{1,2}(S_T)} + h \left\| \frac{\partial y}{\partial x_3} \right\|_{H^1(S_T)} \right). \tag{4.74}$$

For an element T with $T \cap M \neq \emptyset$ we can estimate with Theorem 3.7 for $p = q = 2$ since $W_\beta^{2,2}(\Omega) \hookrightarrow H^1(\Omega)$ (see (2.4))

$$\|y - E_h y\|_{H^1(T)} \leq c \sum_{i=1}^3 h_{i,T} h_{1,T}^{-\beta} \left\| \frac{\partial y}{\partial x_i} \right\|_{W_\beta^{1,2}(T)}$$

$$\leq c \left(\sum_{i=1}^2 h^{(1-\beta)/\mu} \left\| \frac{\partial y}{\partial x_i} \right\|_{W_\beta^{1,2}(T)} + h_{3,T} \left\| \frac{\partial y}{\partial x_3} \right\|_{H^1(T)} \right) \tag{4.75}$$

$$\leq c \left(\sum_{i=1}^2 h \left\| \frac{\partial y}{\partial x_i} \right\|_{W_\beta^{1,2}(T)} + h \left\| \frac{\partial y}{\partial x_3} \right\|_{H^1(T)} \right), \tag{4.76}$$

where we have used the additional regularity of y in x_3-direction (see Corollary 4.18), $r^\beta \leq ch_{1,T}^\beta$ in (4.75) and $\beta = 1 - \mu$.

The estimates (4.74) and (4.76) yield together with the fact that the number of elements in S_T is bounded by a constant the inequality

$$\|y - E_h y\|_{H^1(\Omega)}^2 = \sum_{T \in \mathcal{T}_h} \|y - E_h y\|_{H^1(T)}^2 \leq ch^2 \left(\sum_{i=1}^2 \left\| \frac{\partial y}{\partial x_i} \right\|_{W_\beta^{1,2}(T)}^2 + \left\| \frac{\partial y}{\partial x_3} \right\|_{H^1(T)}^2 \right).$$

Together with the regularity results from Lemma 4.16 and Corollary 4.18 this proves the desired estimate (4.73). $\qquad\square$

With the global interpolation error estimates from Theorem 4.21 and 4.22 at hand, we are able to prove the finite element error estimates in the following two theorems with standard techniques.

Theorem 4.23. *Let y be the solution of (4.45) and let y_h be the finite element solution defined by (4.64). Assume that the mesh fulfills condition (2.13) with $\mu < \pi/\omega$. Then the finite element error can be estimated by*

$$|y - y_h|_{H^1(\Omega)} \leq ch\|f\|_{L^2(\Omega)}, \tag{4.77}$$

$$\|y - y_h\|_{L^2(\Omega)} \leq ch^2\|f\|_{L^2(\Omega)}. \tag{4.78}$$

Proof. Since $V_{0h} \subset V_0$ it follows from (4.45) and (4.64) that

$$a(y - y_h, v_h) = 0 \qquad \text{for all } v_h \in V_{0h}.$$

Since $E_{0h}y \in V_{0h}$ we can conclude

$$\|\nabla(y - y_h)\|_{L^2(\Omega)}^2 \leq c \cdot a(y - y_h, y - y_h) = c \cdot a(y - y_h, y - E_{0h}y)$$

$$\leq c\|\nabla(y - y_h)\|_{L^2(\Omega)} \cdot \|\nabla(y - E_{0h}y)\|_{L^2(\Omega)}$$

what results with the interpolation error estimate (4.66) in

$$\|\nabla(y - y_h)\|_{L^2(\Omega)} \leq c\|\nabla(y - E_{0h}y)\|_{L^2(\Omega)} \leq ch\|f\|_{L^2(\Omega)}$$

and (4.77) is shown. For the proof of (4.78) we use the Aubin-Nitsche trick: Let $w \in V_0$ be the solution of

$$a_D(v, w) = (y - y_h, v) \quad \forall v \in V_0$$

and w_h the corresponding finite element solution. In analogy to (4.77) one has

$$|w - w_h|_{H^1(\Omega)} \leq ch\|y - y_h\|_{L^2(\Omega)}.$$

Consequently, this yields

$$\|y - y_h\|_{L^2(\Omega)}^2 = a_D(y - y_h, w)$$

$$= a_D(y - y_h, w - w_h)$$

$$\leq c|y - y_h|_{H^1(\Omega)}|w - w_h|_{H^1(\Omega)}$$

$$\leq ch^2\|f\|_{L^2(\Omega)}\|y - y_h\|_{L^2(\Omega)}.$$

Division by $\|y - y_h\|_{L^2(\Omega)}$ yields assertion (4.78). □

Remark 4.24. In the proof of Theorem 4.23 it was essential that $E_{0h}y \in V_{0h}$ for $y \in V_0$, i.e., that the interpolation operator preserves homogeneous boundary conditions. This was the reason for introducing E_{0h} in Chapter 3.

Theorem 4.25. *Let y be the solution of (4.47) and let y_h be the finite element solution defined by (4.65). Assume that the mesh fulfills condition (2.13) with $\mu < \pi/\omega$. Then the finite element error can be estimated by*

$$|y - y_h|_{H^1(\Omega)} \leq ch\|f\|_{L^2(\Omega)}$$

$$\|y - y_h\|_{L^2(\Omega)} \leq ch^2\|f\|_{L^2(\Omega)}.$$

Proof. The H^1-estimate follows from inequality (4.73) like the assertion (4.77) in Theorem 4.23 from (4.66). The application of the Aubin-Nitsche trick yields again the L^2-estimate. □

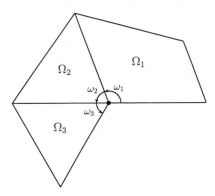

Figure 4.2: Example for subdomains Ω_i in interface problem

4.3 Scalar elliptic equations with nonsmooth coefficients

In this section we consider the interface problem for the Laplacian. Such problems arise from diffusive processes when different materials are involved and therefore the diffusion coefficient varies and admits discontinuities. In this thesis we do not want to treat interface problems in the most general setting but we rather want to show that the type of singularities that occur there can be handled with similar techniques as singularities caused by nonsmooth domains. The same type of mesh grading as for elliptic equations in corner domains leads optimal convergence rates for the finite element error. Later in Section 5.2.3 we will show that this extends to the corresponding optimal control problem.

We assume that the domain Ω can be partitioned in disjoint, open, polygonal Lipschitz subdomains Ω_i, $i = 1, \ldots, n$, on which the diffusion coefficient k has the constant value k_i. Since the singular behaviour is a local phenomenon we restrict our considerations to one corner located at the origin and assume that no singularities occur at the other corners. The interior angle of the subdomains Ω_i at this corner is denoted by ω_i, see Figure 4.2 for an example with $n = 3$. For $n = 1$ the state equation reduces to the Poisson equation. This case is treated in [15]. With $y_i := y|_{\Omega_i}$, $i = 1, \ldots, n$, the interface problem can be written as

$$
\begin{aligned}
-k_i \Delta y_i &= f && \text{in } \Omega_i, \quad i = 1, \ldots, n, \\
y_i(r, \omega_i) &= y_{i+1}(r, \omega_i) && \qquad\quad i = 1, \ldots, n-1, \\
k_i \frac{\partial y_i(r, \omega_i)}{\partial \varphi} &= k_{i+1} \frac{\partial y_{i+1}(r, \omega_i)}{\partial \varphi} && \qquad\quad i = 1, \ldots, n-1, \\
y &= 0 && \text{on } \partial\Omega.
\end{aligned}
\tag{4.79}
$$

The variational formulation reads as

$$\text{Find } y \in V_0: \qquad a_I(y,v) = (f,v)_{L^2(\Omega)} \qquad \forall v \in V_0 \tag{4.80}$$

with bilinear form $a_I : H^1(\Omega) \times H^1(\Omega) \to \mathbb{R}$,

$$a_I(y,v) := \int_\Omega k \nabla y \cdot \nabla v \tag{4.81}$$

and $V_0 := \{v \in H^1(\Omega) : v|_{\partial\Omega} = 0\}$.

4.3.1 Regularity

From the interface condition in (4.79) one can conclude

$$k_i \frac{\partial y_i}{\partial n_i}|_{\partial\Omega_i \cap \partial\Omega_j} + k_j \frac{\partial y_j}{\partial n_j}|_{\partial\Omega_i \cap \partial\Omega_j} = 0$$

for two adjacent subdomains Ω_i, Ω_j and outward normals to the interface n_i, n_j. Since $k_i \neq k_j$ the normal derivatives have a jump discontinuity across the interface. This means $y \notin H^{3/2}(\Omega)$. Extensive studies on the regularity also for more general settings can be found in [100, 106]. As in the case of elliptic equations with corner singularities we describe the regularity of the solution of (4.80) in weighted Sobolev spaces. To this end, we introduce the *Sturm-Liouville* eigenvalue problem,

$$-\Phi_i''(\varphi) = \lambda^2 \Phi_i(\varphi), \quad \varphi \in (\omega_{i-1}, \omega_i), \quad i = 1, \dots, n \tag{4.82}$$

with the boundary and interface conditions

$$\Phi_1(0) = \Phi_n(\omega) = 0 \qquad \qquad \text{(boundary conditions)}$$
$$\Phi_i(\omega_i) = \Phi_{i+1}(\omega_i) \quad i = 1, \dots, n-1 \qquad \text{(interface conditions)}$$
$$k_i \Phi_i'(\omega_i) = k_{i+1} \Phi_{i+1}'(\omega_i) \quad i = 1, \dots, n-1. \qquad \text{(interface conditions)}$$

Lemma 4.26. *Let λ be the smallest positive solution of (4.82), $p \in (0,\infty)$, $\beta > 2 - 2/p - \lambda$ and $f \in \mathcal{V}_\beta^{0,p}(\Omega)$. Then there exists a unique solution $y \in \mathcal{V}_\beta^{2,p}(\Omega)$ of (4.80) and the inequality*

$$\|y\|_{\mathcal{V}_\beta^{2,p}(\Omega)} \leq c \|f\|_{\mathcal{V}_\beta^{0,p}(\Omega)}$$

holds.

Proof. The assertion follows from [100, Theorem 3.6] where it is proved in a more general setting in a cone. In Example 2.29, p. 102 of that book is shown that λ has to solve the eigenvalue problem (4.82). $\qquad \square$

Remark 4.27. According to [100, Theorem 2.27] the solution of (4.80) can be decomposed into a regular part $y_r \in \mathcal{W}^{2,p}(\Omega)$ and a singular part y_s, i.e., $y = y_s + y_r$, as long as $f \in L^p(\Omega)$. In Example 2.29 of that book the exact form of y_s is given such that one can conclude $y_s \in \mathcal{V}_\beta^{2,p}(\Omega)$. This is also true for more general boundary conditions. Notice, that as a consequence of the Hardy inequalities $y_r \in \mathcal{W}^{2,p}(\Omega)$ implies $y_r \in \mathcal{V}_\beta^{2,p}(\Omega)$, $\beta \geq 0$ in case of homogeneous Dirichlet boundary conditions, see [100, Theorem 1.19].

Remark 4.28. For a numerical evaluation of λ for different ω_i and k_i we refer to Figure 2.3–2.5 in [100]. One realizes that $\lambda < 1/2$ is possible in contrast to the case of corner singularities of elliptic problems (comp. Subsection 4.1.1).

4.3.2 Finite element error estimates

We discretize the boundary value problem (4.80) by a finite element scheme. With an admissible triangulation $\mathcal{T}_h$, see page 16, and

$$V_{0h} = \{v_h \in C(\bar{\Omega}) : v_h|_T \in \mathcal{P}_1 \text{ for all } T \in \mathcal{T}_h \text{ and } v_h = 0 \text{ on } \partial\Omega\} \tag{4.83}$$

the discretized problem is formulated as

$$\text{Find } y_h \in V_{0h} : \qquad a_I(y_h, v_h) = (f, v_h)_{L^2(\Omega)} \qquad \forall v_h \in V_{0h}. \tag{4.84}$$

We assume that the underlying triangulation $\mathcal{T}_h$ is aligned with the partition of Ω, i.e., the boundary $\partial\Omega_i$ is made up of edges of triangles in T_h.

There are many papers around concerning finite element error estimates for interface problems. Let us mention here at least the early works of Babuška [21], Baker [24] and King [76]. Different treatments of smooth interfaces were studied e.g. in [26] and [33].

Theorem 4.29. *Let y and y_h be the solution of (4.80) and (4.84), respectively. On a mesh of type (2.12) with grading parameter $\mu < \lambda$ the estimate*

$$\|y - y_h\|_{L^2(\Omega)} + h\|y - y_h\|_{H^1(\Omega)} \leq ch^2 |y|_{\mathcal{V}_\beta^{2,2}(\Omega)} \leq ch^2 \|f\|_{L^2(\Omega)}$$

is valid for $\beta > 1 - \lambda$.

Proof. According to Lemma 4.26 one has $y \in \mathcal{V}_\beta^{2,2}(\Omega)$ for $\beta > 1 - \lambda$. An estimate of the interpolation error for functions of $\mathcal{V}_\beta^{2,2}(\Omega)$ can be proved by standard arguments using the piecewise linear Lagrange interpolant, see, e.g.,[22]. By Céa's Lemma this yields directly the H^1-estimate. The L^2-estimate follows by the Aubin-Nitsche method. $\square$

4.4 Stokes equations in nonsmooth domains

The aim of this section is to prove finite element error estimates for the Stokes equations in nonsmooth domains. To this end we first consider a rather general situation and prove error bounds under certain assumptions on the regularity of the solution and for discrete approximation spaces that fulfill some reasonable conditions. In the subsections 4.4.2 and 4.4.3 we apply these results for particular domains and approximation spaces.

4.4.1 General situation

We consider the boundary value problem

$$
\begin{aligned}
-\Delta v + \nabla q &= f && \text{in } \Omega, \\
\nabla \cdot v &= 0 && \text{in } \Omega, \\
v &= 0 && \text{on } \Gamma = \partial\Omega.
\end{aligned}
\tag{4.85}
$$

Here, Ω is a bounded subset of $\mathbb{R}^d$, $d = 2, 3$ with boundary $\partial\Omega$. These equations, also known as Stokes equations, describe the flow of an incompressible and viscous fluid in a $d-$dimensional body. We call $v : \Omega \to \mathbb{R}^d$ *velocity field* and $q : \Omega \to \mathbb{R}$ *pressure*. The pressure is determined by (4.85) only up to an additive constant. Therefore one usually considers a normalized solution, i.e. the condition

$$
\int_\Omega q = 0
$$

is satisfied.

The variational formulation is given as the saddle point problem

$$
\begin{aligned}
\text{Find } (v, q) &\in X \times M : \\
a(v, \varphi) + b(\varphi, q) &= (f, \varphi) && \forall \varphi \in X \\
b(v, \psi) &= 0 && \forall \psi \in M
\end{aligned}
\tag{4.86}
$$

with the bilinear forms $a : X \times X \to \mathbb{R}$ and $b : X \times M \to \mathbb{R}$ defined as

$$
a(v, \varphi) := \sum_{i=1}^d \int_\Omega \nabla v_i \cdot \nabla \varphi_i \quad \text{and} \quad b(\varphi, q) := -\int_\Omega q \nabla \cdot \varphi,
$$

and the spaces

$$
X = \left\{ v \in (H^1(\Omega))^d : v|_{\partial\Omega} = 0 \right\} \quad \text{and} \quad M = \left\{ q \in L^2(\Omega) : \int_\Omega q = 0 \right\}.
$$

Notice, that the problem (4.86) admits a unique solution $(v, q) \in X \times M$, see e.g. [61, Th. I.5.1].

As we have seen for scalar elliptic equations, it is convenient to describe the regularity of the solution in weighted Sobolev spaces as soon as one likes to include nonconvex domains Ω. Since we may consider problems with corner- or edge-singularities, we introduce the general weighted Sobolev spaces $H^k_\omega(\Omega)^d$, $k = 1, 2$. The corresponding norm is defined as

$$
\|v\|_{H^k_\omega(\Omega)^d} = \left(\sum_{|\alpha| \le k} \|\omega_\alpha D^\alpha v\|^2_{L^2(\Omega)^d} \right)^{1/2}
$$

where ω_α is a suitable positive weight depending on the concrete problem under consideration. Later in our particular examples, this weight will be determined by the distance to the singular points. In order to stay in a more general setting, we do not specify the weights for the moment and state the following assumption.

Assumption FE1. For the solution of the Stokes problem one has for a sufficiently smooth right-hand side f

$$(v, q) \in H^2_{\omega_1}(\Omega)^d \times H^1_{\omega_2}(\Omega)$$

with suitable weights ω_1 and ω_2.

Since we like to have a continuous velocity field, we formulate the next assumption.

Assumption FE2. The embedding $H^2_\omega(\Omega) \hookrightarrow C(\bar{\Omega})$ holds.

For the numerical solution of the saddle point problem (4.86) we choose a velocity approximation space X_h and a pressure approximation space M_h each consisting of piecewise polynomial functions, such that $M_h \subset M$ but not necessarily $X_h \subset X$. Additionally we assume

$$X_h \subset \{v_h \in L^2(\Omega)^d : v_h|_T \in H^1(T)^d \ \forall T \in \mathcal{T}_h\}.$$

Since the velocity space X may not include the discrete velocity space X_h, we define the approximate solution of (4.86) by using the weaker bilinear forms $a_h : X_h \times X_h \to \mathbb{R}$ and $b_h : X_h \times M_h \to \mathbb{R}$ with

$$a_h(v_h, \varphi_h) := \sum_{T \in \mathcal{T}_h} \sum_{i=1}^{d} \int_T \nabla v_{h,i} \cdot \nabla \varphi_{h,i} \quad \text{and} \quad b_h(\varphi_h, p_h) := -\sum_{T \in \mathcal{T}_h} \int_T p_h \nabla \cdot \varphi_h.$$

$$(4.87)$$

Here, the i-th component of the vectors v_h and φ_h is denoted by $v_{h,i}$ and $\varphi_{h,i}$, respectively. The bilinear form $a_h(\cdot, \cdot)$ induces a broken $H^1(\Omega)$-norm by $\|\cdot\|_{X_h} := a_h(\cdot, \cdot)^{1/2}$. With this definitions at hand we can formulate the finite element approximation by

Find $(v_h, q_h) \in X_h \times M_h$ such that
$$a_h(v_h, \varphi_h) + b_h(\varphi_h, q_h) = (f, \varphi_h) \qquad \forall \varphi_h \in X_h \qquad (4.88)$$
$$b_h(v_h, \psi_h) = 0 \qquad \forall \psi_h \in M_h. \qquad (4.89)$$

In the following we formulate a couple of assumptions that will together with Assumptions FE1 and FE2 be sufficient to prove finite element error estimates.

Assumption FE3. There exist interpolation operators $i_h^v : H^2_\omega(\Omega)^d \cap X \to X_h \cap X$ and $i_h^p : H^1_\omega(\Omega) \cap M \to M_h$ such that for the solution $(v, q) \in X \times M$ of the Stokes problem (4.86) the interpolation properties

(i) $\|v - i_h^v v\|_{X_h} \le ch\|v\|_{H^2(\Omega)} \le ch\|f\|_{L^2(\Omega)^d}$

(ii) $\|v - i_h^v v\|_{L^\infty(\Omega)} \le c\|f\|_{L^2(\Omega)^d}$

(iii) $\|q - i_h^p q\|_{L^2(\Omega)} \le ch\|q\|_{H^1_\omega(\Omega)} \le ch\|f\|_{L^2(\Omega)^d}$

hold.

Assumption FE4. There exists a p satisfying

$$p < \infty \quad \text{if } d = 2,$$
$$p \le 6 \quad \text{if } d = 3, \tag{4.90}$$

such that the inverse estimate

$$\|\varphi_h\|_{L^\infty(\Omega)} \le ch^{-1}\|\varphi_h\|_{L^p(\Omega)} \qquad \forall \varphi_h \in X_h$$

is valid.

Assumption FE5. A consistency error estimate holds for the space X_h,

$$|a_h(v, \varphi_h) + b_h(\varphi_h, q) - (f, \varphi_h)| \le ch\|\varphi_h\|_{X_h}\|f\|_{L^2(\Omega)} \qquad \forall (f, \varphi_h) \in L^2(\Omega) \times X_h.$$

where $(v, q) \in X \times M$ is the solution of the Stokes problem (4.86).

Assumption FE6. The pair (X_h, M_h) fulfills the uniform discrete inf-sup-condition, i.e. there exists a positive constant β independent of h such that

$$\inf_{\psi_h \in M_h} \sup_{\varphi_h \in X_h} \frac{b(\varphi_h, \psi_h)}{\|\varphi_h\|_{X_h}\|\psi_h\|_M} \ge \beta.$$

Theorem 4.30. *Assume that Assumptions FE1 through FE6 hold. Let (v, q) be the solution of (4.86) and let (v_h, q_h) be the finite element solution defined by (4.88)–(4.89). Then the approximation error can be estimated by*

$$\|q - q_h\|_{L^2(\Omega)} + \|v - v_h\|_{X_h} \le ch\|f\|_{L^2(\Omega)^d} \tag{4.91}$$
$$\|v - v_h\|_{L^2(\Omega)^d} \le ch^2\|f\|_{L^2(\Omega)^d} \tag{4.92}$$
$$\|v - v_h\|_{L^p(\Omega)^d} \le ch\|f\|_{L^2(\Omega)^d} \text{ for } p \text{ satisfying (4.90)} \tag{4.93}$$
$$\|v - v_h\|_{L^\infty(\Omega)^d} \le c\|f\|_{L^2(\Omega)^d} \tag{4.94}$$

Proof. From [35, Proposition II.2.16] one has

$$\|v - v_h\|_{X_h} + \|q - q_h\|_{L^2(\Omega)} \le c \inf_{\varphi_h \in X_h} \|v - \varphi_h\|_{X_h} + c \inf_{\mu_h \in M_h} \|q - \mu_h\|_{L^2(\Omega)}$$
$$+ c \sup_{\varphi_h \in X_h} \frac{|a_h(v, \varphi_h) + b_h(\varphi_h, q) - (f, v_h)|}{\|\varphi_h\|_{X_h}}$$

Then estimate (4.91) can be concluded from Assumptions FE3 and FE5.

In order to prove the L^2-error estimate (4.92) we apply a non-conforming version of the Aubin-Nitsche method. Therefore we consider for $g \in L^2(\Omega)^d$ the solution $(\varphi_g, \psi_g) \in (H_0^1(\Omega)^d \cap H_\omega^2(\Omega)^d) \times (L^2(\Omega) \cap H_\omega^1(\Omega))$ of the saddle-point problem

$$a(\varphi, \varphi_g) - b(\varphi, \psi_g) = (g, \varphi) \qquad \forall \varphi \in X$$
$$b(\varphi_g, \psi) = 0 \qquad \forall \psi \in M. \tag{4.95}$$

We introduce for $(\varphi, q, v) \in X \times M \times X$ the abbreviations

$$d_{1,h}(\varphi, q, v) := a_h(\varphi, v) + b_h(v, q) - (f, v),$$
$$d_{1,h}(\varphi, q, v) := a_h(v, \varphi) - b_h(v, q) - (g, v).$$

Then one has for $\varphi_h \in X_h$ and $\psi_h \in M_h$

$$
\begin{aligned}
&a_h(v - v_h, \varphi_g - \varphi_h) - b_h(v - v_h, \psi_g - \psi_h) + b_h(\varphi_g - \varphi_h, q - q_h) \\
&\quad - d_{1,h}(v, q, \varphi_g - \varphi_h) - d_{2,h}(\varphi_g, \psi_g, v - v_h) \\
&= -a_h(v - v_h, \varphi_h) + b_h(v - v_h, \psi_h) - b_h(\varphi_g - \varphi_h, q_h) - \\
&\qquad a_h(v, \varphi_g) + a_h(v, \varphi_h) + (f, \varphi_g - \varphi_h) + (g, v - v_h) \\
&= \quad a_h(v_h, \varphi_h) - (f, \varphi_h) - a_h(v, \varphi_g) + (f, \varphi_g) - \\
&\qquad b_h(\varphi_g - \varphi_h, q_h) - b_h(v - v_h, \psi_h) + (g, v - v_h) \\
&= -b_h(\varphi_h, q_h) + b_h(\varphi_g, q) - b_h(\varphi_g - \varphi_h, q_h) + b_h(v - v_h, \psi_h) + (g, v - v_h) \\
&= -b_h(\varphi_g, q_h) - b_h(v_h, \psi_h) + b_h(v, \psi_h) + (g, v - v_h) \\
&= \quad (g, v - v_h)
\end{aligned}
$$

In the last two steps we have used (4.95) and (4.89), respectively. Furthermore, one has $b_h(v, \psi_h) = 0$ since $M_h \subset M$. Now we can continue with

$$
\begin{aligned}
\|v - v_h\|_{L^2(\Omega)^d} &= \sup_{0 \neq g \in L^2(\Omega)^d} \frac{(g, v - v_h)}{\|g\|_{L^2(\Omega)^d}} \\
&\leq \sup_{0 \neq g \in L^2(\Omega)^d} \|g\|_{L^2(\Omega)^d}^{-1} \left(|a_h(v - v_h, \varphi_g - \varphi_h)| + |b_h(v - v_h, \psi_g - \psi_h)| + \right. \\
&\qquad + |b_h(\varphi_g - \varphi_h, q - q_h)| + |d_{1,h}(v, q, \varphi_g - \varphi_h)| + |d_{2,h}(\varphi_g, \psi_g, v - v_h)| \big) .
\end{aligned}
$$
$$(4.96)$$

We estimate these terms separately. For the first term we set $\varphi_h = i_h^v \varphi_g$, where i_h^v is the interpolation operator of Assumption FE3. This yields

$$
\begin{aligned}
|a_h(v - v_h, \varphi_g - i_h^v \varphi_g)| &\leq c \|v - v_h\|_{X_h} \|\varphi_g - i_h^v \varphi_g\|_{X_h} \\
&\leq ch^2 \|f\|_{L^2(\Omega)^d} \|g\|_{L^2(\Omega)^d}.
\end{aligned}
$$
$$(4.97)$$

To estimate the second term we set $\psi_h = i_h^p \psi_g$ with the operator i_h^p of Assumption FE3. Then one has

$$
\begin{aligned}
|b_h(v - v_h, \psi_g - i_h^p \psi_g)| &\leq \|v - v_h\|_{X_h} \|\psi_g - i_h^p \psi_g\|_{L^2(\Omega)} \\
&\leq ch^2 \|f\|_{L^2(\Omega)^d} \|g\|_{L^2(\Omega)^d}.
\end{aligned}
$$
$$(4.98)$$

where we have used (4.91) and Assumption FE3 (ii). The third term can be estimated by

$$
\begin{aligned}
|b_h(\varphi_g - i_h^v \varphi_g, q - q_h)| &\leq \|\varphi_g - i_h^v \varphi_g\|_{X_h} \|q - q_h\|_{L^2(\Omega)} \\
&\leq ch^2 \|f\|_{L^2(\Omega)^d} \|g\|_{L^2(\Omega)^d}
\end{aligned}
$$
$$(4.99)$$

where we used the properties of i_h^v given in Assumption FE3 and the L^2-error estimate for q in (4.91). Since $\varphi_g - i_h^v \varphi_g \in X$ there holds for the fourth term

$$d_{1,h}(v, q, \varphi_g - i_h^v \varphi_g) = a_h(v, \varphi_g - i_h^v \varphi_g) + b_h(\varphi_g - i_h^v \varphi_g, q) - (f, \varphi_g - i_h^v \varphi_g) = 0.$$
(4.100)

Finally, the fifth term yields

$$
\begin{aligned}
|d_{2,h}(\varphi_g, \psi_g, v - v_h)| &= |a_h(\varphi_g, v - v_h) + b_h(v - v_h, \psi_g) - (g, v - v_h)| \\
&\leq |a_h(\varphi_g, v - i_h^v v) + b_h(v - i_h^v v, \psi_g) - (g, v - i_h^v v)| \\
&\quad + |a_h(\varphi_g, i_h^v v - v_h) + b_h(i_h^v v - v_h, \psi_g) + (g, i_h^v v - v_h)|.
\end{aligned}
$$
(4.101)

Since $v - i_h^v v \in X$ we can conclude like above

$$|a_h(\varphi_g, v - i_h^v v) + b_h(v - i_h^v v, \psi_g) - (g, v - i_h^v v)| = 0.$$
(4.102)

The consistency error estimate of Assumption FE5 entails

$$|a_h(\varphi_g, i_h^v v - v_h) + b_h(i_h^v v - v_h, \psi_g) + (g, i_h^v v - v_h)| \leq ch \|i_h^v v - v_h\|_{X_h} \|g\|_{L^2(\Omega)}.$$
(4.103)

With equations (4.102) and (4.103) we can continue from (4.101) with

$$
\begin{aligned}
|d_h(\varphi_g, \psi_g, v - v_h)| &\leq ch \|i_h^v v - v_h\|_{X_h} \|g\|_{L^2(\Omega)^d} \\
&\leq ch (\|v - v_h\|_{X_h} + \|v - i_h^v v\|_{X_h}) \|g\|_{L^2(\Omega)^d} \\
&\leq ch^2 \|f\|_{L^2(\Omega)^d} \|g\|_{L^2(\Omega)^d}
\end{aligned}
$$

where we have used again (4.91) and Assumption FE3 (i). This last estimate implies together with (4.96)–(4.100) the assertion (4.92).

Estimate (4.93) follows directly from inequality (4.91) by the embedding $H^1(\Omega) \hookrightarrow L^p(\Omega)$ for p satisfying (4.90).

In order to prove inequality (4.94), we can conclude from the triangle inequality and Assumptions FE3 and FE4

$$
\begin{aligned}
\|v - v_h\|_{L^\infty(\Omega)^d} &\leq \|v - i_h^v v\|_{L^\infty(\Omega)^d} + \|v_h - i_h^v v\|_{L^\infty(\Omega)^d} \\
&\leq c\|f\|_{L^2(\Omega)^d} + ch^{-1} \|v_h - i_h^v v\|_{L^p(\Omega)^d} \\
&\leq c\|f\|_{L^2(\Omega)^d} + ch^{-1} \left(\|v - v_h\|_{L^p(\Omega)^d} + \|v - i_h^v v\|_{L^p(\Omega)^d} \right)
\end{aligned}
$$
(4.104)

for a certain p satisfying (4.90). Since $H^1(\Omega) \hookrightarrow L^p(\Omega)$ for such a p one can conclude from Assumption FE3 (i)

$$\|v - i_h^v v\|_{L^p(\Omega)^d} \leq ch \|f\|_{L^2(\Omega)^d}.$$

With this estimate we can continue from (4.104) and get together with inequality (4.93) the desired result (4.94). □

4.4.2 Stokes equations in polygonal domains

We consider the Stokes system (4.85),

$$
\begin{aligned}
-\Delta v + \nabla q &= f && \text{in } \Omega, \\
\nabla \cdot v &= 0 && \text{in } \Omega, \\
v &= 0 && \text{on } \Gamma = \partial\Omega.
\end{aligned}
\tag{4.105}
$$

Here, Ω is a bounded, polygonal subset of $\mathbb{R}^2$. Again, we restrict our considerations to the case of only one reentrant corner in the domain located at the origin. More general settings can be reduced to this situation by a localization argument.

4.4.2.1 Regularity

According to the results of [46] the vertex singularity of the solution (v, q) of (4.105) is related to a pole λ of the inverse of an operator, which results from writing the Stokes system in spherical coordinates centered at the vertex and applying an integral transformation. It turns out that λ is the smallest positive solution of a certain eigenvalue problem. We specify this in the following lemma.

Lemma 4.31. *Assume that $f \in L^p(\Omega)^2$, $2 \le p < \infty$ and let $\lambda > 0$ be the smallest positive solution of*

$$
\sin(\lambda\omega) = -\lambda \sin\omega,
\tag{4.106}
$$

where ω is the interior angle at the corner. Then the solution $(v, q) \in X \times M$ of the Stokes problem (4.105) satisfies

$$
v \in V_\beta^{2,p}(\Omega)^2 \text{ and } q \in V_\beta^{1,p}(\Omega) \quad \forall \beta > 2 - \lambda - \frac{2}{p}
$$

and the a priori estimate

$$
\|v\|_{V_\beta^{2,p}(\Omega)^2} + \|q\|_{V_\beta^{1,p}(\Omega)} \le c\|f\|_{L^p(\Omega)^2}
$$

holds.

Proof. The assertion follows from [79, Theorem 5.8.1]. If one sets $\alpha = \omega$ and $\xi = \lambda\omega$ in that theorem, one directly admits the desired regularity result. $\qquad\square$

Remark 4.32. The smallest positive solution λ of (4.106) satisfies

$$
\frac{1}{2} < \lambda < \frac{\pi}{\omega},
$$

see e.g. [46].

Remark 4.33. In the setting of (4.105) the spaces $V_\beta^{k,2}(\Omega)$ play the role of the general spaces $H_\omega^k(\Omega)$. With $\omega_{1\alpha} = \omega_{2\alpha} = r^{\beta-k+\alpha}$ Assumption FE1 is satisfied.

4.4.2.2 Finite element error estimate

Now we concentrate on error estimates for the finite element approximation (v_h, q_h) as introduced in (4.88)–(4.89) for the particular setting of this subsection. In order to counteract the corner singularity at the origin, we discretize the domain Ω by a family of graded triangulations as introduced in Subsection 2.3.1. We require the grading parameter μ to satisfy $\mu < \lambda$. With this grading parameter at hand, we are able to prove Assumption FE2 for this situation.

Proof of Assumption FE2. Since $\mu < \lambda$ it is $1 - \mu > 1 - \lambda$ and it follows from Lemma 4.31 with $\beta = 1 - \mu$ that $(v, q) \in V_{1-\mu}^{2,2}(\Omega) \times V_{1-\mu}^{2,1}(\Omega)$. This means Assumption FE2 is satisfied due to the embedding

$$V_{1-\mu}^{2,2}(\Omega) \hookrightarrow V_0^{2-(1-\mu),2}(\Omega) \hookrightarrow H^{1+\mu}(\Omega) \hookrightarrow C(\bar{\Omega}). \tag{4.107}$$

The first embedding is proved in [113, Lemma 1.2]. The second one follows directly from the definition of the spaces. The last embedding is a conclusion from the Sobolev embedding theorem. □

In the following we give examples of pairs of spaces (X_h, M_h) that satisfy the Assumptions FE3 through FE6.

Conforming elements We first concentrate on the case $X_h \subset X$ and $M_h \subset M$. An overview can be found e.g. in [61].

a) Bernardi-Raugel-Fortin element

$$X_h = \{v_h \in H_0^1(\Omega)^2 : v_h|_T \in \mathcal{P}_1^+ \ \forall T \in \mathcal{T}_h\}$$
$$M_h = \{q_h \in L_0^2(\Omega) : q_h|_T \in \mathcal{P}_0 \ \forall T \in \mathcal{T}_h\}$$

where $\mathcal{P}_1^+ = \mathcal{P}_1^2 \oplus \mathrm{span}\{n_1\lambda_2\lambda_3, n_2\lambda_3\lambda_1, n_3\lambda_1\lambda_2\}$

b) $(\mathcal{P}_2, \mathcal{P}_0)$

c) Mini-element

$$X_h = \{v_h \in H_0^1(\Omega)^2 : v_h|_T \in \mathcal{P}_1^+ \ \forall T \in \mathcal{T}_h\}$$
$$M_h = \{q_h \in C(\bar{\Omega}) \cap L_0^2(\Omega) : q_h|_T \in \mathcal{P}_1 \ \forall T \in \mathcal{T}_h\}$$

where $\mathcal{P}_1^+ = [\mathcal{P}_1^2 \oplus \mathrm{span}\{\lambda_1\lambda_2\lambda_3\}]^2$

d) Taylor-Hood element $(\mathcal{P}_k, \mathcal{P}_{k-1})$, $k \geq 2$

We check Assumptions FE3 through FE6 for these cases and summarize the error estimate in Lemma 4.34.

Proof of Assumption FE3. We use as interpolation operator i_h^v the standard Lagrange interpolant I_h. This is possible due to the fact that piecewise linear and continuous functions are contained in the space X_h. Although the proof of (i) follows by standard arguments, we sketch it here. We write

$$\|\nabla(v - I_h v)\|^2_{L^2(\Omega)} = \sum_{T \in \mathcal{T}_h} \|\nabla(v - I_h v)\|^2_{L^2(T)}.$$

In case of $r_T > 0$ we can use the H^2-regularity in T and conclude

$$\|\nabla(v - I_h v)\|^2_{L^2(T)} \leq ch^2_T |v|^2_{H^2(T)} \leq ch^2_T r_T^{-2\beta} |u|^2_{V^{2,2}_\beta(T)}.$$

With $\beta = 1 - \mu$ one has from (2.12) the equality $h^2_T r_T^{-2\beta} = ch^2 r_T^{2-2\mu} r_T^{-2+2\mu} = ch^2$. This yields

$$\|\nabla(v - I_h v)\|^2_{L^2(T)} \leq ch^2 |u|^2_{V^{2,2}_\beta(T)}.$$

For $r_T = 0$ we use the triangle inequality to estimate

$$\|\nabla(v - I_h v)\|^2_{L^2(T)} \leq c \left(\|\nabla v\|^2_{L^2(T)} + \|\nabla(I_h v)\|^2_{L^2(T)} \right)$$

We estimate an the two terms separately. For the first term we get

$$\|\nabla v\|^2_{L^2(T)} = \|r^{1-\beta} r^{\beta-1} \nabla v\|^2_{L^2(T)} \leq ch^{2(1-\beta)}_T \|r^{\beta-1} \nabla v\|^2_{L^2(T)} \leq ch^{2(1-\beta)}_T \|v\|^2_{V^{2,2}_\beta(T)}.$$

The second term can be estimated using a inverse inequality and the embedding $V^{2,2}_\beta(\Omega) \hookrightarrow L^\infty(\Omega)$,

$$\|\nabla I_h v\|_{L^2(T)} \leq ch^{-1}_T \|I_h v\|_{L^2(T)}$$
$$\leq ch^{-1}_T |T|^{1/2} \|\hat{I}_h v\|_{L^2(\hat{T})}$$
$$\leq ch^{-1}_T |T|^{1/2} \|v\|_{L^\infty(\hat{T})}$$
$$\leq ch^{-1}_T |T|^{1/2} \|v\|_{V^{2,2}_\beta(\hat{T})}$$
$$\leq ch^{1-\beta}_T \|v\|_{V^{2,2}_\beta(T)}.$$

According to (2.12) one can conclude for $\beta = 1 - \mu$ the equation $h^{1-\beta}_T = h_T^{\frac{1-(1-\mu)}{\mu}} = ch$ and therefore

$$\|\nabla(v - I_h v)\|^2_{L^2(T)} \leq ch^2 |v|^2_{V^{2,2}_\beta(T)}$$

in the case $r_T = 0$. Summing up over all elements yields finally

$$\|\nabla(v - I_h v)\|_{L^2(\Omega)} \leq ch \|u\|^2_{V^{2,2}_{1-\mu}(\Omega)} \leq ch \|f\|_{L^2(\Omega)}$$

where we have used the a priori estimate of Lemma 4.31 with $\beta = 1 - \mu > 1 - \lambda$. This proves (i).

For the proof of estimate (ii) we distinguish again between elements with $r_T = 0$ and $r_T > 0$. In the case of $r_T = 0$ we start with the triangle inequality and get

$$\|v - I_h v\|_{L^\infty(T)} \leq \|v\|_{L^\infty(T)} + \|I_h v\|_{L^\infty(T)}. \tag{4.108}$$

For the first term of the right-hand side it follows

$$\|v\|_{L^\infty(T)} \leq \|\hat{v}\|_{L^\infty(\hat{T})} \leq c\|\hat{v}\|_{V_\beta^{2,2}(\hat{T})} \leq h_T^{1-\beta}\|v\|_{V_\beta^{2,2}(T)}.$$

Since $\|I_h v\|_{L^\infty(T)} \leq \|v\|_{L^\infty(T)}$ this estimate is also valid for the second term. With the same argumentation as above inequality (4.108) yields with $\beta = 1 - \mu$ for elements with $r_T = 0$

$$\|v - I_h v\|_{L^\infty(T)} \leq ch\|v\|_{V_{1-\mu}^{2,2}(T)}. \tag{4.109}$$

For $r_T > 0$ we use again the H^2-regularity in T and conclude

$$\begin{aligned}\|v - I_h v\|_{L^\infty(T)} &\leq ch_T^2 |T|^{-1/2}|v|_{H^2(T)} = ch_T|v|_{H^2(T)} \\ &\leq chr_T^{1-\mu}|v|_{H^2(T)} \leq ch|v|_{V_{1-\mu}^{2,2}(T)}\end{aligned} \tag{4.110}$$

If one denotes by T^* the element where $v - I_h v$ attaines its maximum, one can follow from inequalities (4.109) and (4.110) together with the a priori estimate of Lemma 4.31

$$\|v - I_h v\|_{L^\infty(\Omega)} = \|v - I_h v\|_{L^\infty(T^*)} \leq ch\|v\|_{V_{1-\mu}^{2,2}(T^*)} \leq ch\|v\|_{V_{1-\mu}^{2,2}(\Omega)} \leq ch\|f\|_{L^2(\Omega)}.$$

This is assumption (ii).

For i_h^p we choose the $L^2(\Omega)$-projection in the space of piecewise constant functions. Then (iii) can be proved similarly to (i). $\qquad\square$

Proof of Assumption FE4. For $\varphi_h \in X_h$ one has

$$\|\varphi_h\|_{L^\infty(T)} = \|\hat{\varphi}_h\|_{L^\infty(\hat{T})} \leq c\|\hat{\varphi}_h\|_{L^p(\hat{T})} = c|T|^{-1/p}\|\varphi_h\|_{L^p(T)}.$$

We choose the smallest element size $h_T = h^{1/\mu}$ and $p \geq \frac{2}{\mu}$. This yields

$$\|\varphi_h\|_{L^\infty(T)} \leq ch^{-\frac{2}{\mu p}}\|\varphi_h\|_{L^p(T)} \leq ch^{-1}\|\varphi_h\|_{L^p(T)},$$

what proves the assertion. $\qquad\square$

Proof of Assumption FE5. Since the discretizations a)–d) are all conforming the consistency error estimate is trivially satisfied. $\qquad\square$

Proof of Assumption FE6. For the proof of the inf-sup-condition for the above element pairs we refer to [61]. $\qquad\square$

Lemma 4.34. *The finite element solution of* (4.105) *satisfies the estimates of Theorem 4.30 for the element pairs described in a)–d) on finite element meshes of type* (2.12) *with grading parameter* $\mu < \lambda$.

Nonconforming element In this paragraph we investigate a discretization X_h of the velocity space X with $X_h \not\subset X$, namely the Crouzeix-Raviart finite element space,

$$X_h := \left\{ v_h \in L^2(\Omega)^2 : v_h|_T \in (\mathcal{P}_1)^2 \ \forall T, \int_E [v_h]_E = 0 \ \forall E \right\} \tag{4.111}$$

where E denotes an edge of an element and $[v_h]_E$ means the jump of v_h on the edge E,

$$
[v_h(x)]_E := \begin{cases} \lim_{\alpha \to 0} (v_h(x + \alpha n_E) - v_h(x - \alpha n_E)) & \text{for an interior edge E,} \\ v_h(x) & \text{for a boundary edge E.} \end{cases}
$$

Here n_E is the outer normal of E. For the approximation of the pressure we use piecewise constant functions, this means

$$
M_h := \left\{ q_h \in L^2(\Omega) : q_h|_T \in \mathcal{P}_0 \ \forall T, \int_\Omega q_h = 0 \right\}. \tag{4.112}
$$

Since Assumptions FE1 and FE2 are checked in Subsection 4.4.2.1 and at the beginning of this subsection it remains to check the Assumptions FE3 through FE6.

Proof of Assumptions FE3–FE4. The proofs of these Assumptions are exactly the same as for the conforming elements a)–d). $\qquad\square$

Proof of Assumption FE5. In the following we choose $\mu > 1/2$. According to Remark 4.32 there exist values for μ with $1/2 < \mu < \lambda$. For the proof of the consistency error estimate we follow the ideas of [32, III.1], where the assertion is proved for quasi-uniform meshes and the Poisson equation in convex domains. For $\varphi_h \in X_h$ we can compute

$$
\begin{aligned}
|a_h(v, \varphi_h) + b_h(\varphi_h, q) - (f, \varphi_h)| &= \left| \sum_T \left(\int_T \nabla \cdot v \nabla \varphi_h \, dx - \int_T q \nabla \cdot \varphi_h \, dx - \int_T f \varphi_h \, dx \right) \right| \\
&= \left| \sum_T \left(\int_{\partial T} \partial_n v \cdot \varphi_h \, ds - \int_T \Delta v \varphi_h \, dx + \int_T \nabla q \cdot \varphi_h \, dx \right. \right. \\
&\qquad \left. \left. + \int_{\partial T} q \varphi_h \cdot n \, ds - \int_T f \varphi_h \, dx \right) \right| \\
&\leq \left| \sum_T \int_{\partial T} \partial_n v \cdot \varphi_h \, ds \right| + \left| \sum_T \int_{\partial T} q \varphi_h \cdot n \, ds \right|. \tag{4.113}
\end{aligned}
$$

In the last step we have used $-\Delta v + \nabla q = f$ in $H^{-1}(\Omega)$ and the triangle inequality. We estimate the two terms in (4.113) separately. In the first term every inner edge occurs two times, but with different signs of the normal derivatives. Therefore the value is not changing, if one adds the constant $\overline{\varphi_h(E)} := 1/|E| \int_E \varphi_h \, dx$ on every edge E. This yields with the Lagrange interpolant I_h

$$
\begin{aligned}
\sum_T \int_{\partial T} \partial_n v \cdot \varphi_h \, ds &= \sum_T \sum_{E \in \partial T} \int_E \partial_n v \cdot \left(\varphi_h - \overline{\varphi_h(E)} \right) ds \\
&= \sum_T \sum_{E \in \partial T} \int_E \partial_n (v - I_h v) \cdot \left(\varphi_h - \overline{\varphi_h(E)} \right) ds.
\end{aligned}
$$

In the last step we utilized the fact that $\int_E \left(\varphi_h - \overline{\varphi_h(E)} \right) ds = 0$ and that $\partial_n I_h v$ is constant. With the Cauchy-Schwarz inequality we end up with

$$\sum_T \int_{\partial T} \partial_n v \cdot \varphi_h \, ds \leq \sum_T \sum_{E \in \partial T} \|\nabla(v - I_h v)\|_{L^2(E)^2} \|\varphi_h - \overline{\varphi_h(E)}\|_{L^2(E)^2}. \tag{4.114}$$

For estimating the right-hand side of (4.114) we distinguish elements at the corner and those far from the corner. Let us first consider elements T with $r_T > 0$. Since $v \in H^2(T)$ we can conclude with the embedding $H^1(\Omega) \hookrightarrow L^2(\partial\Omega)$ (comp. Theorem 2.14) and the Bramble-Hilbert-Lemma

$$\|\nabla(\hat{v} - \hat{I}\hat{v})\|_{L^2(\partial\hat{T})^2}^2 \leq c\|\hat{v} - \hat{I}\hat{v}\|_{H^2(\hat{T})^2}^2 \leq c|\hat{v}|_{H^2(\hat{T})^2}^2.$$

Transformation on T yields

$$\|\nabla(v - I_h v)\|_{L^2(E)^2}^2 \leq ch_T |v|_{H^2(T)^2}^2.$$

With the same argumentation one can conclude

$$\|\varphi_h - \overline{\varphi_h(E)}\|_{L^2(E)^2}^2 \leq ch_T |\varphi_h|_{H^1(T)^2}^2. \tag{4.115}$$

These two estimates yield with (2.12)

$$\|\nabla(v - I_h v)\|_{L^2(E)^2}^2 \|\varphi_h - \overline{\varphi_h(E)}\|_{L^2(E)^2}^2 \leq ch^2 r_T^{2-2\mu} |v|_{H^2(T)^2}^2 |\varphi_h|_{H^1(T)^2}^2$$

$$\leq ch^2 |v|_{V_{1-\mu}^{2,2}(T)^2}^2 |\varphi_h|_{H^1(T)^2}^2. \tag{4.116}$$

Now we consider elements T with $r_T = 0$. The embedding $H^\mu(\Omega) \hookrightarrow L^2(\partial\Omega)$ for $\mu > 1/2$ (comp. Theorem 2.14) results in

$$\|\nabla(v - I_h v)\|_{L^2(E)^2} \leq ch_T^{-1} \cdot |E|^{1/2} \|\nabla(v - Iv)\|_{L^2(\hat{E})^2} \leq ch_T^{-1/2} \|v - Iv\|_{H^{1+\mu}(\hat{T})^2}.$$

The embedding $V_{1-\mu}^{2,2}(\Omega) \hookrightarrow H^{1+\mu}(\Omega)$ (comp. (4.107)) and the boundedness of the nodal interpolation operator I result in

$$\|\nabla(v - I_h v)\|_{L^2(E)^2} \leq ch_T^{-1/2} \|v\|_{V_{1-\mu}^{2,2}(\hat{T})^2} \leq ch_T^{\mu-1/2} \|v\|_{V_{1-\mu}^{2,2}(T)^2}.$$

Like above one has

$$\|\varphi_h - \overline{\varphi_h(E)}\|_{L^2(E)^2}^2 \leq ch_T |\varphi_h|_{H^1(T)^2}^2, \tag{4.117}$$

such that we can conclude for elements T with $r_T = 0$ with (2.12)

$$\|\nabla(v - I_h v)\|_{L^2(E)^2} \|\varphi_h - \overline{\varphi_h(E)}\|_{L^2(E)^2} \leq ch \|v\|_{V_{1-\mu}^{2,2}(T)^2} |\varphi_h|_{H^1(T)^2}. \tag{4.118}$$

Summing up over all elements we get from (4.114), (4.116) and (4.118) the estimate

$$\left| \sum_T \int_{\partial T} \partial_n v \cdot \varphi_h \, ds \right| \leq ch \|v\|_{V_{1-\mu}^{2,2}(\Omega)^2} \|\varphi_h\|_{X_h} \tag{4.119}$$

for the first term of the right-hand side of (4.113). For the second term of (4.113) we proceed very similar. With the L^2-projection Q_h in the space of piecewise constant functions,

$$Q_h q(x) = \frac{1}{|T|} \int_T q \, dx \quad \text{for } x \in T,$$

we can write with the same argumentation as above

$$\left| \sum_T \int_{\partial T} q \varphi_h \cdot n \, ds \right| \leq \sum_T \sum_{E \in \partial T} \|q - Q_h q\|_{L^2(E)} \|\varphi_h - \overline{\varphi_h(E)}\|_{L^2(E)^2}. \tag{4.120}$$

For elements T with $r_T > 0$ the embedding $H^1(\Omega) \hookrightarrow L^2(\partial\Omega)$ (comp. Theorem 2.14) yields together with the Bramble-Hilbert-Lemma and (2.12)

$$\|q - Q_h q\|_{L^2(E)}^2 \leq c h_T |q|_{H^1(T)}^2$$

Taking (4.115) and (2.12) into account it follows

$$\|q - Q_h q\|_{L^2(E)} \|\varphi_h - \overline{\varphi_h(E)}\|_{L^2(E)^2} \leq c h^2 r_T^{2-2\mu} |q|_{H^1(T)}^2 \|\varphi_h\|_{H^1(T)^2}^2$$
$$\leq c h |q|_{V_{1-\mu}^{1,2}(T)}^2 \|\varphi_h\|_{H^1(T)^2}^2. \tag{4.121}$$

For elements T with $r_T = 0$ we use the embedding $H^\mu(\Omega) \hookrightarrow L^2(\partial\Omega)$ and the boundedness of Q_h to conclude

$$\|q - Q_h q\|_{L^2(E)} < c|E|^{1/2} \|\hat{q} - \hat{Q}\hat{q}\|_{L^2(\hat{F})} \leq c h_T^{1/2} \|\hat{q}\|_{H^\mu(\hat{T})}$$

The embedding $V_{1-\mu}^{1,2}(\Omega) \hookrightarrow H^\mu(\Omega)$ (comp. (4.107)) yields

$$\|q - Q_h q\|_{L^2(E)} \leq c h_T^{1/2} \|\hat{q}\|_{V_{1-\mu}^{1,2}(\hat{T})} \leq h_T^{\mu-1/2} \|q\|_{V_{1-\mu}^{1,2}(T)}$$

Together with (4.117) and (2.12) we end up with

$$\|q - Q_h q\|_{L^2(E)} \|\varphi_h - \overline{\varphi_h(E)}\|_{L^2(E)^2} \leq c h_T^\mu \|q\|_{V_{1-\mu}^{1,2}(T)} \|\varphi_h\|_{H^1(T)^2}$$
$$= c h \|q\|_{V_{1-\mu}^{1,2}(T)} \|\varphi_h\|_{H^1(T)^2}. \tag{4.122}$$

If one sums up over all elements one obtains with the estimates (4.120) – (4.122)

$$\left| \sum_T \int_{\partial T} q \varphi_h \cdot n \, ds \right| \leq c h \|q\|_{V_{1-\mu}^{1,2}(\Omega)} \|\varphi_h\|_{X_h}. \tag{4.123}$$

The estimates (4.113), (4.119) and (4.123) prove together with the regularity results in Lemma 4.31 the assertion. $\quad\square$

Proof of Assumption FE6. For an anisotropic discretization on prismatic domains this assertion is shown in [13, Lemma 3.1]. The same standard arguments apply in our twodimensional setting on isotropic graded meshes. For the sake of completeness we repeat them here. First of all, we introduce the Crouzeix-Raviart interpolant $\Pi_{cr} : X \to X_h$ which is defined elementwise by

$$\int_E v = \int_E \Pi_{cr} v \quad \forall E \in \partial T, \ \forall T \in \mathcal{T}_h.$$

Since $\Pi_{cr} v$ is linear on every element T and $\frac{\partial(\Pi_{cr} v)}{\partial n_T}$ is constant on every edge of an element T it follows by Green's formula

$$|\Pi_{cr} v|^2_{H^1(T)} = \int_{\partial T} \frac{\partial(\Pi_{cr} v)}{\partial n_T} \cdot \Pi_{cr} v \, ds = \int_{\partial T} \frac{\partial(\Pi_{cr} v)}{\partial n_T} \cdot v \, ds = \int_T \nabla(\Pi_{cr} v) \nabla v \, dx.$$

The application of the Cauchy-Schwarz inequality and division by $|\Pi_{cr} v|_{H^1(T)}$ yield the stability estimate

$$|\Pi_{cr} v|_{H^1(T)} \le |v|_{H^1(T)}. \tag{4.124}$$

We consider an arbitrary but fixed $\psi_h \in M_h$. According to [61, Corollary I.2.4] there exists $\varphi \in X$, satisfying

$$\nabla \cdot \varphi = \psi_h, \quad |\varphi|_{H^1(\Omega)} \le c\|\psi_h\|_{L^2(\Omega)}. \tag{4.125}$$

With the stability estimate (4.124) and Green's formula one can conclude

$$\int_T \nabla \cdot \varphi \, dx = \sum_{E \in \partial T} \int_E \varphi \cdot n \, ds = \sum_{E \in \partial T} \int_E \Pi_{cr} \varphi \cdot n \, ds = \int_T \nabla \cdot \Pi_{cr} \varphi \, dx.$$

This equation results in

$$b_h(\Pi_{cr}\varphi, \psi_h) = -\sum_T \int_T \psi_h \nabla \cdot \Pi_{cr}\varphi \, dx = -\sum_T \int_T \psi_h \nabla \cdot \varphi \, dx = \|\psi_h\|^2_{L^2(\Omega)}, \tag{4.126}$$

where we have utilized the fact that $\psi_h|_T$ is constant and the properties (4.125) of ψ_h. Furthermore, we obtain by combination of (4.124) and (4.125) the estimate

$$\|\Pi_{cr}\varphi\|_{X_h} \le c|\varphi|_{H^1(\Omega)} \le c\|\psi_h\|_{L^2(\Omega)}.$$

This last inequality yields together with (4.126)

$$\sup_{\varphi_h \in X_h} \frac{b_h(\varphi_h, \psi_h)}{\|\varphi_h\|_{X_h}\|\psi_h\|_{L^2(\Omega)}} \ge \frac{b_h(\Pi_{cr}\varphi, \psi_h)}{\|\Pi_{cr}\varphi\|_{X_h}\|\psi_h\|_{L^2(\Omega)}} \ge c.$$

This proves the assertion since ψ_h was chosen arbitrarily. $\qquad\square$

We checked all the assumptions such that we can summarize our result in the following lemma.

Lemma 4.35. *The finite element solution of (4.105) satisfies the estimates of Theorem 4.30 for the element pair (X_h, M_h) defined in (4.111) and (4.112) on finite element meshes of type (2.12) with grading parameter $1/2 < \mu < \lambda$.*

4.4.3 Stokes equations in prismatic domains

In this subsection we consider again the boundary value problem (4.85)

$$
\begin{aligned}
-\Delta v + \nabla q &= f & \text{in } \Omega, \\
\nabla \cdot v &= 0 & \text{in } \Omega, \\
v &= 0 & \text{on } \Gamma = \partial\Omega,
\end{aligned}
\tag{4.127}
$$

but now in a prismatic domain as defined for the boundary value problem (4.43), i.e. $\Omega = G \times Z \subset \mathbb{R}^3$ with a bounded polygonal domain $G \subset R^2$ and a interval $Z := (0, z_0) \subset \mathbb{R}$. Again, we assume that G has only one corner with interior angle $\omega > \pi$ at the origin.

4.4.3.1 Regularity

In the following lemma we prove the regularity in the same weighted spaces as we did for the scalar elliptic Dirichlet problem (comp. Lemma 4.15). Furthermore one can also observe some extra regularity in edge direction. Notice, that in contrast to Lemma 4.17 the results for the regularity in edge direction of solutions of the Stokes equations cover only the case $p = 2$.

Lemma 4.36. *Assume that $f \in L^p(\Omega)^3$, $1 < p < \infty$ and let $\lambda > 0$ be the smallest positive solution of*

$$
\sin(\lambda\omega) = -\lambda \sin\omega
\tag{4.128}
$$

where ω is the interior angle at the edge. Then the solution $(v, q) \in X \times M$ of the Stokes problem (4.127) satisfies

$$
v \in V_\beta^{2,p}(\Omega)^3 \text{ and } q \in V_\beta^{1,p}(\Omega) \quad \forall \beta > 2 - \lambda - \frac{2}{p}
\tag{4.129}
$$

and the a priori estimate

$$
\|v\|_{V_\beta^{2,p}(\Omega)^3} + \|q\|_{V_\beta^{1,p}(\Omega)} \leq c\|f\|_{L^p(\Omega)}
\tag{4.130}
$$

holds. Further one has for $f \in L^2(\Omega)^3$

$$
\partial_3 v \in V_0^{1,2}(\Omega)^3 \quad \text{and} \quad \partial_3 q \in L^2(\Omega)
\tag{4.131}
$$

with the corresponding a priori estimate

$$
\|\partial_3 v\|_{V_0^{1,2}(\Omega)^3} + \|\partial_3 q\|_{L^2(\Omega)} \leq c\|f\|_{L^2(\Omega)}.
\tag{4.132}
$$

Proof. The assertions (4.129) and (4.130) follow from Theorem 6.1 of [89]. In our case for the vertex eigenvalues λ_q the inequality $\text{Re}\lambda_q \geq 1$ holds [101]. This means we can choose $\beta_q = 0$ in Theorem 6.1 of [89]. So setting $m = 2$ in this theorem yields (4.129) and (4.130). The extra regularity in edge direction stated in (4.131) and (4.132) is proved in Theorem 2.1 of [13]. □

Remark 4.37. The leading singularity of u_3 can be characterized by $r^{\pi/\omega}$ [13]. On the other hand the smallest positive solution λ of (4.128) satisfies

$$\frac{1}{2} < \lambda < \frac{\pi}{\omega},$$

see e.g. [46], such that the global regularity is dominated by r^λ.

Remark 4.38. In the setting of (4.127) the spaces $V_\beta^{k,2}(\Omega)$ play the role of the general spaces $H_\omega^k(\Omega)$. With $\omega_{1\alpha} = \omega_{2\alpha} = r^{\beta-k+\alpha}$ Assumption FE1 is satisfied.

4.4.3.2 Finite element error estimate

In order to get an optimal rate of convergence for the finite element error, we counteract the edge singularity which results from the reentrant edge in the domain Ω by an anisotropic graded mesh. In detail, we use meshes as introduced in Subsection 2.3.2 with grading parameter $1/2 < \mu < \lambda$. Due to Remark 4.37 such a μ exists. We approximate the velocity by Crouzeix-Raviart elements,

$$X_h := \left\{ v_h \in L^2(\Omega)^3 : v_h|_T \in (\mathcal{P}_1)^3 \ \forall T, \int_F [v_h]_F = 0 \ \forall F \right\} \tag{4.133}$$

where F denotes a face of an element and $[v_h]_F$ means the jump of v_h on the face F,

$$[v_h(x)]_F := \begin{cases} \lim_{\alpha \to 0} (v_h(x + \alpha n_F) - v_h(x - \alpha n_F)) & \text{for an interior face F,} \\ v_h(x) & \text{for a boundary face F.} \end{cases}$$

Here n_F is the outer normal of F. For the approximation of the pressure we use piecewise constant functions, this means

$$M_h := \left\{ q_h \in L^2(\Omega) : q_h|_T \in \mathcal{P}_0 \ \forall T, \int_\Omega q_h = 0 \right\}. \tag{4.134}$$

For the proof of the finite element error estimates we check the assumptions stated in Subsection 4.4.1. Since Assumption FE1 is already proved in the foregoing paragraph, we continue with Assumptions FE2 through FE6.

Proof of Assumption FE2. This proof is very similar to the one for the two dimensional case. Since $\mu < \lambda$ it is $1 - \mu > 1 - \lambda$ and we can choose $\beta = 1 - \mu$ in Lemma 4.36. This yields $(v, q) \in V_{1-\mu}^{2,2}(\Omega) \times V_{1-\mu}^{2,1}(\Omega)$. Then the assumption is a consequence of the embedding

$$V_{1-\mu}^{2,2}(\Omega) \hookrightarrow V_0^{2-(1-\mu),2}(\Omega) \hookrightarrow W^{1+\mu,2}(\Omega) \hookrightarrow C(\bar{\Omega}). \tag{4.135}$$

The first embedding follows from [113, Lemma 1.2]. The second one can be concluded directly from the definition of the spaces. The last embedding is a conclusion from the Sobolev embedding theorem and the fact that $\mu > 1/2$ and therefore $1 + \mu - \frac{3}{2} > 0$. $\quad\square$

Proof of Assumption FE3. As interpolant in the velocity space we choose $i_h^v := E_{0h}$ with E_{0h} defined in (3.5). Notice that we do not use the Crouzeix-Raviart interpolant although we use Crouzeix-Raviart elements for the velocity and although the estimates in Assumption FE3 could be fulfilled by this interpolant. The reason for this is that the Crouzeix-Raviart interpolant maps to X_h but not to $X \cap X_h$ as demanded in this assumption. The proof of estimate (i) is the same as the one for Theorem 4.21. One just has to repeat the arguments there componentwise and to plugin the regularity results from Lemma 4.36. In order to prove (ii) we write

$$E_{0h}v(x) = \sum_{i \in I} \left[\frac{1}{|\sigma_i|} \int_{\sigma_i} v \right] \varphi_i(x).$$

Since v is according to Assumption FE2 a continuous function, we can conclude

$$\|E_{0h}v\|_{L^\infty(\Omega)} = \sup_{i \in I} \left| \frac{1}{|\sigma_i|} \int_{\sigma_i} v \right| \leq \|v\|_{L^\infty(\bar\Omega)}. \tag{4.136}$$

The embedding $V_{1-\mu}^{2,2}(\Omega) \hookrightarrow L^\infty(\Omega)$, see (4.135), yields together with Lemma 4.36

$$\|v\|_{L^\infty(\bar\Omega)} \leq c\|v\|_{V_{1-\mu}^{2,2}(\Omega)} \leq \|f\|_{L^2(\Omega)}. \tag{4.137}$$

Now we can conclude with the triangle inequality and the estimates (4.136) and (4.137)

$$\|v - E_{0h}v\|_{L^\infty(\Omega)} \leq \|v\|_{L^\infty(\Omega)} + \|E_{0h}v\|_{L^\infty(\Omega)} \leq \|f\|_{L^2(\Omega)}$$

and (ii) is proved. For the proof of (iii) we set $i_h^p := Q_h$ with Q_h being the $L^2(\Omega)$-projection in the space of piecewise constant functions, i.e.

$$Q_h q(x) := \frac{1}{|T|} \int_T q(x)\, \mathrm{d}x \quad \text{for } x \in T.$$

The assertion is shown in the proof of Lemma 3.2 in [13]. Notice, that M_h in that proof is the interpolant Q_h in our setting. For the sake of completeness, we sketch the details here. We write the error elementwise,

$$\|q - Q_h q\|_{L^2(\Omega)}^2 = \sum_{T \in \mathcal{T}_h} \|q - Q_h q\|_{L^2(T)}^2. \tag{4.138}$$

For elements T with $r_T > 0$, one has according to [12, (3.5), (3.6)]

$$\|q - Q_h q\|_{L^2(T)} \leq c \sum_{i=1}^{3} h_{i,T} \|\partial_i q\|_{L^2(T)}.$$

This can be continued by

$$\|q - Q_h q\|_{L^2(T)} \le c \sum_{i=1}^{3} h_{i,T} \|\partial_i q\|_{L^2(T)}$$

$$\le c \left(\sum_{i=1}^{2} h_{i,T} r_T^{\mu-1} \|\partial_i q\|_{V_{1-\mu}^{0,2}(T)} + h_{3,T} \|\partial_3 q\|_{L^2(T)} \right)$$

$$\le c \left(h \sum_{i=1}^{2} \|\partial_i q\|_{V_{1-\mu}^{0,2}(T)} + h \|\partial_3 q\|_{L^2(T)} \right), \qquad (4.139)$$

where we have used the regularity results of Lemma 4.36 and the definition of the mesh sizes (2.13). For elements T with $r_T = 0$ we use the fact that Q_h is bounded and therefore

$$\|q - Q_h q\|_{L^2(T)} \le c\|q\|_{L^2(T)} \le \|r^\mu\|_{L^\infty(T)} \|r^{-\mu} q\|_{L^2(T)}$$

$$\le h_{1,T}^\mu \|r^{-\mu} q\|_{L^2(T)} \le ch\|q\|_{V_{1-\mu}^{1,2}(T)}. \qquad (4.140)$$

The estimates (4.138)–(4.140) yield together with Lemma 4.36

$$\|q - Q_h q\|_{L^2(\Omega)} \le ch \left(\|q\|_{V_{1-\mu}^{1,2}(\Omega)} + \|\partial_3 q\|_{L^2(\Omega)} \right) \le ch\|f\|_{L^2(\Omega)}.$$

and Assumption FE3 (iii) is proved. □

Proof of Assumption FE4. For an arbitrary $\varphi_h \in X_h$ one has

$$\|\varphi_h\|_{L^\infty(T)} = \|\hat\varphi_h\|_{L^\infty(\hat T)} \le c\|\hat\varphi_h\|_{L^p(\hat T)} \le c|T|^{-1/p}\|\varphi_h\|_{L^p(T)}.$$

We choose the minimal element size according to (2.13) and arrive at

$$\|\varphi_h\|_{L^\infty(\Omega)} \le ch^{-2/(\mu p)} h^{-1/p} \|\varphi_h\|_{L^p(\Omega)}.$$

In order to achieve $-\frac{2}{\mu p} - \frac{1}{p} \ge -1$ one has to demand $p \ge \frac{2}{\mu} + 1$. This condition is no contradiction to $p \le 6$ as long as $\mu \ge \frac{2}{5}$. But this can be satisfied since $\mu > 1/2$. Therefore it exists $p \in \left[\frac{2}{\mu} + 1, 6 \right]$ such that

$$\|\varphi_h\|_{L^\infty(\Omega)} \le ch^{-1}\|\varphi_h\|_{L^p(\Omega)}.$$

what is the inequality of Assumption FE4. □

Proof of Assumptions FE5 and FE6. The consistency error estimate and the discrete inf-sup-condition are proved in [13]. □

We checked Assumptions FE1 through FE6 such that we can summarize our result in the following lemma.

Lemma 4.39. . *The finite element solution of (4.127) satisfies the estimates of Theorem 4.30 for the element pair (X_h, M_h) defined in (4.133) and (4.134) on finite element meshes of type (2.13) with grading parameter $1/2 < \mu < \lambda$.*

CHAPTER 5

Error estimates for PDE-constrained Optimal Control Problems

5.1 Analysis for general linear-quadratic optimal control problems

We consider the general linear-quadratic control-constrained optimal control problem

$$\min_{(y,u)\in Y\times U} J(y,u) := \frac{1}{2}\|y - y_d\|_Z^2 + \frac{\nu}{2}\|u\|_U^2, \tag{5.1}$$
$$\text{subject to} \quad y = Su, \quad u \in U^{\text{ad}},$$

where Z, $U = U^*$ are Hilbert spaces and Y is a Banach space with $Y \hookrightarrow Z \hookrightarrow Y^*$. We introduce a Banach space $X \hookrightarrow Z$ and demand $y_d \in X$. The operator $S : U \to Y \subset U$ is a linear, bounded control-to-state solution operator. We assume ν to be a fixed positive number and $U^{\text{ad}} \subset U$ nonempty, convex and closed.

Remark 5.1. Later we will choose for S the solution operators of the boundary value problems introduced in Chapter 4. Depending on the problem one can set, e.g., $U = Z = L^2(\Omega)$, $Y = H^1(\Omega)$ and $X = C^{0,\sigma}(\bar{\Omega})$.

Remark 5.2. Problem (5.1) is equivalent to the reduced problem

$$\min_{u\in U^{\text{ad}}} \hat{J}(u) \tag{5.2}$$

with

$$\hat{J}(u) := J(Su, u) = \frac{1}{2}\|Su - y_d\|_Z^2 + \frac{\nu}{2}\|u\|_U^2.$$

5.1.1 Existence and uniqueness of a solution

Definition 5.3. A state-control pair $(\bar{y}, \bar{u}) \in Y \times U^{\mathrm{ad}}$ is called *optimal* for (5.1), if $\bar{y} = S\bar{u}$ and

$$J(\bar{y}, \bar{u}) \leq J(y, u) \quad \forall (y, u) \in Y \times U^{\mathrm{ad}}, \; y = Su.$$

Theorem 5.4. *The optimal control problem (5.1) has a unique optimal solution* $(\bar{y}, \bar{u})$. *Furthermore, for* $S^* : Y^* \to U$ *being the adjoint of* S, *the optimality conditions*

$$\bar{y} = S\bar{u}, \tag{5.3}$$

$$\bar{p} = S^*(S\bar{u} - y_d) \tag{5.4}$$

$$\bar{u} \in U^{\mathrm{ad}}, \quad (\nu\bar{u} + \bar{p}, u - \bar{u})_U \geq 0 \quad \forall u \in U^{\mathrm{ad}} \tag{5.5}$$

hold. These conditions are necessary and sufficient.

Proof. For the proof we consider the reduced form (5.2). The objective functional $\hat{J}$ is strictly convex and radially unbounded, i.e., $\hat{J}(u) \to \infty$ as $\|u\| \to \infty$, $u \in U^{\mathrm{ad}}$. Since U^{ad} is convex and closed, we can apply Theorem 1.3 of [84] and get existence and uniqueness of an optimal solution. The optimality conditions follow from the same theorem by differentiation of $\hat{J}$. The strict convexity implies that the necessary condition is also sufficient. $\qquad \square$

Remark 5.5. In the following we refer to (5.3) as state equation, to (5.4) as adjoint equation and to (5.5) as variational inequality.

Lemma 5.6. *Let* $\Pi_{U^{\mathrm{ad}}} : U \to U^{\mathrm{ad}}$ *be the projection on* U^{ad}, *i.e.,*

$$\Pi_{U^{\mathrm{ad}}}(u) \in U^{\mathrm{ad}}, \quad \|\Pi_{U^{\mathrm{ad}}}(u) - u\|_U = \min_{v \in U^{\mathrm{ad}}} \|v - u\|_U \quad \forall u \in U.$$

Then the projection formula

$$\bar{u} = \Pi_{U^{\mathrm{ad}}}\left(-\frac{1}{\nu}\bar{p}\right) \tag{5.6}$$

is equivalent to the variational inequality in Theorem 5.4.

Proof. The assertion is motivated in [88]. A detailed proof is given, e.g., in [73]. The assertion follows from Lemma 1.11 in that book by setting $\gamma = 1/\nu$. $\qquad \square$

5.1.2 Discretization concepts

5.1.2.1 Variational discrete approach

In [71] Hinze introduced a discretization concept for the optimal control problem (5.2) which is alone based on the discretization of the state space. The control space is not discretized. We introduce the linear, bounded control-to-discretized state operator $S_h : U \to Y_h \subset U$, where Y_h is a finite dimensional subspace equipped with the norm of Y. $S_h^* : Y^* \to Y_h$ denotes the adjoint of S_h. Then the discrete optimal control $\bar{u}_h^s$ is defined via the variational inequality

$$(\nu \bar{u}_h^s + S_h^*(S_h \bar{u}_h^s - y_d), u - \bar{u}_h^s)_U \geq 0 \quad \forall u \in U^{\mathrm{ad}}. \tag{5.7}$$

This variational inequality is actually the necessary and sufficient optimality condition of

$$\min_{u \in U^{\mathrm{ad}}} J_h(u),$$
$$J_h(u) := \frac{1}{2}\|S_h u - y_d\|_Z^2 + \frac{\nu}{2}\|u\|_U^2, \tag{5.8}$$

compare also Theorem 5.4. In order to be able to prove an error estimate the following assumptions are sufficient.

Assumption VAR1. The operators S_h and S_h^* are bounded, i.e., the inequalities

$$\|S_h\|_{U \to U} \leq c \quad \text{and} \quad \|S_h^*\|_{U^* \to U^*} \leq c$$

are valid with a constant c independent of h.

Assumption VAR2. The estimates

$$\|(S - S_h)u\|_U \leq ch^2 \|u\|_U \quad \forall u \in U,$$
$$\|(S^* - S_h^*)z\|_U \leq ch^2 \|z\|_Z \quad \forall z \in Z$$

hold true.

We introduce the optimal discrete state $\bar{y}_h^s := S_h \bar{u}_h^s$ and optimal adjoint state $\bar{p}_h^s := S_h^*(S_h \bar{u}_h^s - y_d)$ and formlate the error estimates in the following theorem.

Theorem 5.7. *Let Assumptions VAR1 and VAR2 hold. Then the estimates*

$$\|\bar{u} - \bar{u}_h^s\|_U \leq ch^2 \left(\|\bar{u}\|_U + \|y_d\|_Z\right), \tag{5.9}$$
$$\|\bar{y} - \bar{y}_h^s\|_U \leq ch^2 \left(\|\bar{u}\|_U + \|y_d\|_Z\right), \tag{5.10}$$
$$\|\bar{p} - \bar{p}_h^s\|_U \leq ch^2 \left(\|\bar{u}\|_U + \|y_d\|_Z\right) \tag{5.11}$$

hold with a constant c independent of h.

Proof. The first estimate is proved in [71]. We recall the arguments here since we have reformulated the assumptions. We use $\bar{u}$ as test function in (5.7) and $\bar{u}_h^s$ as test function in (5.5) and add both inequalities. This yields

$$(\nu(\bar{u} - \bar{u}_h^s) + S^*(S\bar{u} - y_d) - S_h^*(S_h \bar{u}_h^s - y_d), \bar{u}_h^s - \bar{u})_U \geq 0.$$

One can continue with

$$
\begin{aligned}
\nu\|\bar{u} - \bar{u}_h^s\|_U^2 &\leq (S^*S\bar{u} - S_h^*S_h\bar{u}_h^s - (S^* - S_h^*)y_d, \bar{u}_h^s - \bar{u})_U \\
&= ((S^*S - S_h^*S_h)\bar{u} + S_h^*(S_h\bar{u} - S_h\bar{u}_h^s) - (S^* - S_h^*)y_d, \bar{u}_h^s - \bar{u})_U \\
&= ((S^*S - S_h^*S_h)\bar{u} - (S^* - S_h^*)y_d, \bar{u}_h^s - \bar{u})_U - \|S_h\bar{u} - S_h\bar{u}_h^s\|_U^2 \\
&\leq (S^*(S\bar{u} - S_h\bar{u}) + (S^* - S_h^*)S_h\bar{u} - (S^* - S_h^*)y_d, \bar{u}_h^s - \bar{u})_U.
\end{aligned}
$$

The Cauchy-Schwarz inequality and the triangle inequality yield after division by $\|\bar{u} - \bar{u}_h^s\|_U$ the estimate

$$
\nu\|\bar{u} - \bar{u}_h^s\|_U \leq \|S^*\|_{U \to U}\|(S - S_h)\bar{u}\|_U + \|(S^* - S_h^*)(S_h\bar{u})\|_U + \|(S^* - S_h^*)y_d\|_U.
$$

The application of Assumptions VAR1 and VAR2 results together with the boundedness of S^* in the first assertion.

For the proof of the second assertion we write

$$
\begin{aligned}
\|\bar{y} - \bar{y}_h^s\|_U &= \|S\bar{u} - S_h\bar{u}_h^s\|_U \\
&\leq \|S\bar{u} - S_h\bar{u}\|_U + \|S_h(\bar{u} - \bar{u}_h^s)\|_U.
\end{aligned}
$$

Inequality (5.10) follows then from Assumptions VAR1 and VAR2 and (5.9).

For the third assertion we can conclude similarly to above

$$
\begin{aligned}
\|\bar{p}_h - \bar{p}_h^s\|_U &= \|S^*(S u - y_d) - S_h^*(S_h u_h^s - y_d)\|_U \\
&= \|S^*(S\bar{u} - S_h\bar{u}) + (S^* - S_h^*)S_h\bar{u} + S_h^*S_h(\bar{u} - \bar{u}_h^s) - (S^* - S_h^*)y_d\|_U.
\end{aligned}
$$

With the triangle inequality the assertion (5.10) follows from the boundedness of S^*, Assumptions VAR1 and VAR2 and (5.9). $\qquad\square$

Remark 5.8. In [71] a numerical algorithm for the solution of problem (5.8) is given. It is pointed out that for every step of the iteration one has to compute the boundary of the active set exactly. This boundary is not aligned with the mesh and therefore requires the computation and management of additional grid points. Especially in three dimensions and for higher order elements this introduces technical difficulties for the implementation. For this reason we concentrate in this thesis on numerical examples concerning the following postprocessing approach.

5.1.2.2 Postprocessing approach

In this subsection, we consider the reduced problem (5.2). We choose

$$
U = Z = L^2(\Omega)^d, \quad Y = H_0^1(\Omega)^d \text{ or } Y = H^1(\Omega)^d,
$$

where $d \in \{1, 2, 3\}$ depending on the problem under consideration. As space of admissible controls we use

$$
U^{\mathrm{ad}} := \{u \in U : u_a \leq u \leq u_b \text{ a.e.}\},
$$

where $u_a \leq u_b$ are constant vectors from $\mathbb{R}^d$. Then the restriction on the set of continuous functions of the projection in the admissible set reads for a function f as

$$(\Pi_{U^{\text{ad}}} f)(x) := \max(u_a, \min(u_b, f(x)).$$

We assume the existence of a triangulation $\mathcal{T}_h = \{T\}$ of Ω, that is admissible in Ciarlet's sense (comp. page 16). The operators $S_h : U \to Y_h$ and $S_h^* : Y^* \to Y_h$ are finite element approximations of S and S^*, respectively and Y_h is a suitable finite element space. We introduce the discrete control space U_h,

$$U_h = \{u_h \in U : u_h|_T \in (\mathcal{P}_0)^d \text{ for all } T \in \mathcal{T}_h\} \quad \text{and} \quad U_h^{\text{ad}} = U_h \cap U^{\text{ad}}.$$

Then the discretized optimal control problem reads as

$$J_h(\bar{u}_h) = \min_{u_h \in U_h^{\text{ad}}} J_h(u_h),$$

$$J_h(u_h) := \frac{1}{2}\|S_h u_h - y_d\|_{L^2(\Omega)}^2 + \frac{\nu}{2}\|u_h\|_{L^2(\Omega)}^2. \tag{5.12}$$

As in the continuous case, this is a strictly convex and radially unbounded optimal control problem. Consequently, (5.12) admits a unique solution $\bar{u}_h$, that satisfies the necessary and sufficient optimality conditions

$$\bar{y}_h = S_h \bar{u}_h,$$
$$\bar{p}_h = S_h^*(\bar{y}_h - y_d),$$
$$(\nu \bar{u}_h + \bar{p}_h, u_h - \bar{u}_h)_U \geq 0 \quad \forall u_h \in U_h^{\text{ad}}. \tag{5.13}$$

For later use, we introduce the affine operators $Pu = S^*(Su - y_d)$ and $P_h u = S_h^*(S_h u - y_d)$, that maps a given control u to the adjoint state $p = Pu$ and the approximate adjoint state $p_h = P_h u$, respectively.

Now we follow an idea, that goes back to Meyer and Rösch [95], namely to compute an approximate control in a postprocessing step. The control $\tilde{u}_h$ is constructed as projection of the approximate adjoint state in the set of admissible controls,

$$\tilde{u}_h = \Pi_{U^{\text{ad}}}\left(-\frac{1}{\nu}\bar{p}_h\right). \tag{5.14}$$

In the following we would like to formulate rather general assumptions, that allow to prove discretization error estimates for the optimal control problem (5.12). To this end, we first define two projection operators.

Definition 5.9. For continuous functions f we define the projection R_h in the space $\mathcal{P}_0$ of piecewise constant functions by

$$(R_h f)(x) := f(S_T) \text{ if } x \in T \tag{5.15}$$

where S_T denotes the centroid of the element T.

The operator Q_h projects L^2-functions g in the space $\mathcal{P}_0$ of piecewise constant functions,

$$(Q_h g)(x) := \frac{1}{|T|} \int_T g(x) \, dx \text{ for } x \in T. \tag{5.16}$$

Both operators are defined componentwise for vector valued functions.

In the following we formulate four general assumptions, that allow to prove that the approximate solution $\bar{u}_h$ is asymptotically closer (in the L^2-sense) to the interpolant $R_h \bar{u}$ than to the optimal control $\bar{u}$. This was originally discovered by Meyer and Rösch [95] for an optimal control problem governed by the Poisson equation. Such a fact is often referred to as *supercloseness*. Based on this result, we show that the approximate control $\tilde{u}_h$, which is constructed according to (5.14) converges in $L^2(\Omega)$ to the optimal control $\bar{u}$ with order 2. Due to the fact that $\bar{u}$ was originally approximated by piecewise constant functions, this is a *superconvergence* result.

Assumption PP1. The discrete solution operators S_h and S_h^* are bounded,

$$\|S_h\|_{U \to H_h^1(\Omega)^d} \leq c, \qquad \|S_h^*\|_{U \to H_h^1(\Omega)^d} \leq c,$$
$$\|S_h\|_{U \to L^\infty(\Omega)^d} \leq c, \qquad \|S_h^*\|_{U \to L^\infty(\Omega)^d} \leq c,$$

with constants c independent of h and space

$$H_h^1(\Omega)^d := \left\{ v : \Omega \to \mathbb{R}^d \ : \ \sum_{T \in \mathcal{T}h} \|v\|_{H^1(T)^d}^2 < \infty \right\}.$$

Notice that Assumption PP1 implies

$$\|S_h\|_{U \to U} \leq c \quad \text{and} \quad \|S_h^*\|_{U \to U} \leq c$$

by the embedding $L^\infty(\Omega) \hookrightarrow U$.

Assumption PP2. The finite element error estimates

$$\|(S - S_h)u\|_U \leq ch^2 \|u\|_U \quad \forall u \in U,$$
$$\|(S^* - S_h^*)z\|_U \leq ch^2 \|z\|_Z \quad \forall z \in Z$$

hold.

Assumption PP3. The optimal control $\bar{u}$ and the corresponding adjoint state $\bar{p}$ satisfy the inequality

$$\|Q_h \bar{p} - R_h \bar{p}\|_U \leq ch^2 \left(\|\bar{u}\|_X + \|y_d\|_X \right).$$

for a space $X \hookrightarrow U$. In particular, $\bar{p}$ is continuous, such that $R_h \bar{p}$ is well defined.

Assumption PP4. The optimal control $\bar{u}$ is contained in X and for all functions $\varphi_h \in X_h$ the inequality

$$(Q_h \bar{u} - R_h \bar{u}, \varphi_h)_U \leq ch^2 \|\varphi_h\|_{L^\infty(\Omega)^d} \left(\|\bar{u}\|_X + \|y_d\|_X \right)$$

holds. In particular, $\bar{u}$ is continuous, such that $R_h \bar{u}$ is well defined.

First, we recall a property of Q_h that is proved in [19].

Lemma 5.10. *For f, $g \in H^1(T)$ the inequality*

$$(f - Q_h f, g)_{L^2(T)} \leq ch_T^2 |f|_{H^1(T)} |g|_{H^1(T)}$$

is valid.

Now we can prove the following properties of the operator R_h.

Lemma 5.11. *Assume that the Assumptions PP1 and PP4 hold. Then the estimates*

$$\|S_h \bar{u} - S_h R_h \bar{u}\|_U \leq ch^2 \left(\|\bar{u}\|_X + \|y_d\|_X\right) \tag{5.17}$$

$$\|P_h \bar{u} - P_h R_h \bar{u}\|_U \leq ch^2 \left(\|\bar{u}\|_X + \|y_d\|_X\right) \tag{5.18}$$

are valid.

Proof. The following proof is similar to the one given by Apel and Winkler in [19] in the special case of optimal control of the Poisson equation and a discretization with linear finite elements. A proof under assumptions like PP1 and PP4 for the optimal control of the Stokes equation is given in [102]. First of all, we write

$$
\begin{aligned}
\|S_h \bar{u} - S_h R_h \bar{u}\|_U^2 &= (S_h \bar{u} - S_h R_h \bar{u}, S_h \bar{u} - S_h R_h \bar{u})_U \\
&= (S_h(\bar{u} - R_h \bar{u}), (S_h \bar{u} - y_d) - (S_h R_h \bar{u} - y_d))_U \\
&= (\bar{u} - R_h u, P_h u - P_h R_h \bar{u})_U \\
&= (\bar{u} - Q_h \bar{u}, P_h \bar{u} - P_h R_h \bar{u})_U + (Q_h \bar{u} - R_h \bar{u}, P_h \bar{u} - P_h R_h \bar{u})_U. \tag{5.19}
\end{aligned}
$$

We estimate these two terms separately. From Lemma 5.10, one can conclude

$$\sum_{T \in \mathcal{T}_h} (\bar{u} - Q_h \bar{u}, P_h \bar{u} - P_h R_h \bar{u})_{L^2(T)^d} \leq c \sum_{T \in \mathcal{T}_h} h_T^2 |\bar{u}|_{H^1(T)^d} |P_h \bar{u} - P_h R_h \bar{u}|_{H^1(T)^d}$$

$$\leq ch^2 |\bar{u}|_{H^1(\Omega)^d} \left(\sum_{T \in \mathcal{T}_h} |P_h \bar{u} - P_h R_h \bar{u}|_{H^1(T)^d}^2 \right)^{1/2}. \tag{5.20}$$

Since one can write

$$\sum_{T \in \mathcal{T}_h} |P_h \bar{u} - P_h R_h \bar{u}|_{H^1(T)^d}^2 \leq \|S_h^*(S_h \bar{u} - S_h R_h \bar{u})\|_{H_h^1(\Omega)^d}^2$$

$$\leq \|S_h^*\|_{U \to H_h^1(\Omega)^d}^2 \|S_h \bar{u} - S_h R_h \bar{u}\|_U^2$$

it follows with Assumption PP1 and (5.20)

$$(\bar{u} - Q_h \bar{u}, P_h \bar{u} - P_h R_h \bar{u})_U \leq ch^2 |\bar{u}|_{H^1(\Omega)^d} \|S_h \bar{u} - S_h R_h \bar{u}\|_U. \tag{5.21}$$

According to Lemma 5.6 and the projection formula (5.6) the domain Ω splits in two parts, the inactive part $\mathcal{I}$, where $\bar{u} = -\frac{1}{\nu}\bar{p}$ and the active part $\Omega\backslash\mathcal{I}$, where $\bar{u}$ is constant. Since $|\bar{u}|_{H^1(\Omega\backslash\mathcal{I})^d} = 0$ one has

$$|\bar{u}|_{H^1(\Omega)^d} \leq c\|\bar{p}\|_{H^1(\Omega)^d} \leq c\|S^*\|_{U\rightarrow H^1(\Omega)^d}\|S\bar{u} - y_d\|_U \leq c\left(\|\bar{u}\|_X + \|y_d\|_X\right) \qquad (5.22)$$

where we have used the boundedness of S^* and S and the embedding $X \hookrightarrow U$. So we get from (5.21) and (5.22) the estimate

$$(\bar{u} - Q_h\bar{u}, P_h\bar{u} - P_h R_h\bar{u})_U \leq ch^2\left(\|\bar{u}\|_X + \|y_d\|_X\right)\|S_h\bar{u} - S_h R_h\bar{u}\|_U. \qquad (5.23)$$

In order to estimate the second term of equation (5.19), we utilize Assumption PP4 and get

$$\begin{aligned}(Q_h\bar{u} - R_h\bar{u}, P_h\bar{u} - P_h R_h\bar{u})_U &\leq ch^2\left(\|\bar{u}\|_X + \|y_d\|_X\right)\|P_h\bar{u} - P_h R_h\bar{u}\|_{L^\infty(\Omega)^d} \\ &\leq ch^2\left(\|\bar{u}\|_X + \|y_d\|_X\right)\|S_h^*\|_{U\rightarrow L^\infty(\Omega)^d}\|S_h\bar{u} - S_h R_h\bar{u}\|_U \\ &\leq ch^2\left(\|\bar{u}\|_X + \|y_d\|_X\right)\|S_h\bar{u} - S_h R_h\bar{u}\|_U \end{aligned} \qquad (5.24)$$

by applying Assumption PP1 in the last step. Estimates (5.23) and (5.24) yield together with (5.19) the assertion (5.17). For the proof of the second assertion of this Lemma we write

$$\|P_h\bar{u} - P_h R_h\bar{u}\|_U = \|S_h^* S_h\bar{u} - S_h^* S_h R_h\bar{u}\|_U \leq \|S_h^*\|_{U\rightarrow U}\|S_h\bar{u} - S_h R_h\bar{u}\|_U.$$

The application of Assumption PP1 and inequality (5.17) yields assertion (5.18). $\qquad\square$

Lemma 5.12. *The inequality*

$$\nu\|R_h\bar{u} - \bar{u}_h\|_U^2 \leq (R_h\bar{p} - \bar{p}_h, \bar{u}_h - R_h\bar{u})_U \qquad (5.25)$$

holds.

Proof. This lemma was originally proved in [95] and is based on a combination of the variational inequalities (5.5) and (5.13). For the sake of completeness we sketch the proof here which is also given in [15]. From the variational inequality (5.5) we have pointwise a.e.

$$(\nu\bar{u}(x) + \bar{p}(x)) \cdot (u(x) - \bar{u}(x)) \geq 0, \quad \forall u \in U^{\mathrm{ad}}.$$

We apply this formula for the center of gravity S_T of any triangle T and $u = \bar{u}_h$. This is possible due to the continuity of $\bar{u}$, $\bar{p}$ and $\bar{u}_h$ at these points. We arrive at

$$(\bar{p}(S_T) + \nu\bar{u}(S_T)) \cdot (\bar{u}_h(S_T) - \bar{u}(S_T)) \geq 0 \quad \forall T.$$

Due to the definition of R_h this is equivalent to

$$(R_h\bar{p}(S_T) + \nu R_h\bar{u}(S_T)) \cdot (\bar{u}_h(S_T) - R_h\bar{u}(S_T)) \geq 0 \quad \forall T.$$

Integration over T and summing up over all T yield

$$(R_h\bar{p} + \nu R_h\bar{u}, \bar{u}_h - R_h\bar{u})_U \geq 0.$$

Moreover, one can test the optimality condition (5.13) for $\bar{u}_h$ with the function $R_h\bar{u}$ and obtains

$$(\bar{p}_h + \nu\bar{u}_h, R_h\bar{u} - \bar{u}_h)_U \geq 0.$$

Adding the two last inequalities results in

$$(R_h\bar{p} - \bar{p}_h + \nu(R_h\bar{u} - \bar{u}_h), \bar{u}_h - R_h\bar{u})_U \geq 0$$

which is equivalent to formula (5.25). □

Now we are able to prove the following supercloseness result.

Theorem 5.13. *Assume that Assumptions PP1–PP4 hold. Then the inequality*

$$\|\bar{u}_h - R_h\bar{u}\|_U \leq ch^2 \left(\|\bar{u}\|_X + \|y_d\|_X\right)$$

is valid.

Proof. The following proof is similar to the one of Theorem 4.21 of [127] which is given in the context of optimal control of the Poisson equation. We give the details here to illustrate the validity under the Assumptions PP1 – PP4. From Lemma 5.12 we have

$$\begin{aligned}
\nu\|\bar{u}_h - R_h\bar{u}\|_U^2 &\leq (R_h\bar{p} - \bar{p}_h, \bar{u}_h - R_h\bar{u})_U \\
&= (R_h\bar{p} - \bar{p}, \bar{u}_h - R_h\bar{u})_U + (\bar{p} - P_h R_h\bar{u}, \bar{u}_h - R_h\bar{u})_U \\
&\quad + (P_h R_h\bar{u} - \bar{p}_h, \bar{u}_h - R_h\bar{u})_U.
\end{aligned} \tag{5.26}$$

We estimate these three terms separately. For the first term, we use that Q_h is an L^2-projection and get

$$\begin{aligned}
(R_h\bar{p} - \bar{p}, \bar{u}_h - R_h\bar{u})_U &= (R_h\bar{p} - Q_h\bar{p}, \bar{u}_h - R_h\bar{u})_U + (Q_h\bar{p} - \bar{p}, \bar{u}_h - R_h\bar{u})_U \\
&= (R_h\bar{p} - Q_h\bar{p}, \bar{u}_h - R_h\bar{u})_U.
\end{aligned}$$

The Cauchy-Schwarz inequality yields together with Assumption PP3

$$\begin{aligned}
(R_h\bar{p} - \bar{p}, \bar{u}_h - R_h\bar{u})_U &\leq \|R_h\bar{p} - Q_h\bar{p}\|_U \|\bar{u}_h - R_h\bar{u}\|_U \\
&\leq ch^2 \left(\|\bar{u}\|_X + \|y_d\|_X\right) \|\bar{u}_h - R_h\bar{u}\|_U. \tag{5.27}
\end{aligned}$$

For the second term we apply again the Cauchy-Schwarz inequality and use $\bar{p} = P\bar{u}$, so that we arrive at

$$(\bar{p} - P_h R_h\bar{u}, \bar{u}_h - R_h\bar{u})_U \leq \|Pu - P_h R_h\bar{u}\|_U \|\bar{u}_h - R_h\bar{u}\|_U.$$

With Assumptions PP1 and PP2, Lemma 5.11 and the embedding $X \hookrightarrow U$, one can conclude

$$\|P\bar{u} - P_h R_h \bar{u}\|_U \leq \|S_h^*\|_{U \to U} \|S\bar{u} - S_h \bar{u}\|_U + \|S^* y_d - S_h^* y_d\|_U + \|P_h \bar{u} - P_h R_h \bar{u}\|_U$$
$$\leq ch^2 \left(\|\bar{u}\|_X + \|y_d\|_X\right),$$

and therefore

$$(\bar{p} - P_h R_h \bar{u}, u_h - R_h \bar{u})_U \leq ch^2 \left(\|\bar{u}\|_X + \|y_d\|_X\right) \|\bar{u}_h - R_h \bar{u}\|_U. \tag{5.28}$$

The third term can simply be omitted since

$$(P_h R_h \bar{u} - \bar{p}_h, \bar{u}_h - R_h \bar{u})_U = (P_h R_h \bar{u} - P_h \bar{u}_h, \bar{u}_h - R_h \bar{u})_U$$
$$= (S_h(R_h \bar{u} - \bar{u}_h), S_h(\bar{u}_h - R_h \bar{u}))_U$$
$$\leq 0. \tag{5.29}$$

Thus, one can conclude from the estimates (5.26)–(5.29)

$$\nu \|\bar{u}_h - R_h \bar{u}\|_U^2 \leq ch^2 \left(\|\bar{u}\|_X + \|y_d\|_X\right) \|u_h - R_h \bar{u}\|_U$$

what yields the assertion. $\qquad\square$

In the following theorem, we formulate the main result of this subsection.

Theorem 5.14. *Assume that the Assumptions PP1–PP4 hold. Then the estimates*

$$\|\bar{y} - \bar{y}_h\|_U \leq ch^2 \left(\|\bar{u}\|_X + \|y_d\|_X\right), \tag{5.30}$$
$$\|\bar{p} - \bar{p}_h\|_U \leq ch^2 \left(\|\bar{u}\|_X + \|y_d\|_X\right), \tag{5.31}$$
$$\|\bar{u} - \tilde{u}_h\|_U \leq ch^2 \left(\|\bar{u}\|_X + \|y_d\|_X\right) \tag{5.32}$$

are valid with a positive constant c independent of h.

Proof. In order to prove the first assertion we apply the triangle inequality and get

$$\|\bar{y} - \bar{y}_h\|_U = \|S\bar{u} - S_h \bar{u}_h\|_U$$
$$\leq \|Su - S_h u\|_U + \|S_h \bar{u} - S_h R_h \bar{u}\|_U + \|S_h(R_h \bar{u} - \bar{u}_h)\|_U.$$

The first term is a finite element error and is estimated in the first inequality of Assumption PP2. For the second term an upper bound is given in Lemma 5.11. For the third term we use the supercloseness result of Theorem 5.13 and the boundedness of S_h given in Assumption PP1. These three estimates yield assertion (5.30). In a similar way one can prove inequality (5.31). By using the Lipschitz continuity of the projection operator, we get

$$\|\bar{u} - \tilde{u}_h\|_U = \left\| \Pi_{U^{ad}} \left(-\frac{1}{\nu} \bar{p} \right) - \Pi_{U^{ad}} \left(-\frac{1}{\nu} \bar{p}_h \right) \right\|_U \leq \frac{1}{\nu} \|\bar{p} - \bar{p}_h\|_U$$

and inequality (5.32) is a direct consequence of estimate (5.31). $\qquad\square$

5.2 Scalar elliptic state equation

We consider the linear–quadratic optimal control problem (5.2),

$$
\hat{J}(\bar{u}) = \min_{u \in U^{\mathrm{ad}}} \hat{J}(u)
$$

$$
\hat{J}(u) := \frac{1}{2}\|Su - y_d\|_{L^2(\Omega)}^2 + \frac{\nu}{2}\|u\|_{L^2(\Omega)}^2
$$

(5.33)

where the operator S associates the state $y = Su$ to the control u as the weak solution of a boundary value problem with right-hand side u in a polygonal domain (4.1), in a polyhedral domain (4.43) or with discontinuous coefficients (4.79). The desired state y_d is assumed to be Hölder continuous, i.e. $y_d \in C^{0,\sigma}(\bar{\Omega})$, $\sigma \in (0,1)$. Furthermore, the set of admissible controls is

$$
U^{\mathrm{ad}} := \{u \in L^2(\Omega) : u_a \leq u \leq u_b \ \text{a.e. in } \Omega\}.
$$

Moreover we introduce the adjoint problem

$$
L^*p = y - y_d \quad \text{in } \Omega, \qquad B^*p = 0 \quad \text{on } \Gamma
$$

where the operator L^* depends on the state equation,

$$
L^*p := \begin{cases} -\nabla \cdot A(x)\nabla p - a_1(x) \cdot \nabla p + a_0(x)p & \text{for state equation (4.1),} \\ -\Delta p & \text{for state equation (4.44),} \\ -\Delta p + p & \text{for state equation (4.46).} \end{cases}
$$

The operator B^* is given as

$$
B^*p = \begin{cases} p & \text{for state equations (4.1) and (4.44),} \\ \frac{\partial p}{\partial n} & \text{for state equation (4.46).} \end{cases}
$$

For the state equation (4.79) we have with $L_i^* = L^*|_{\Omega_i}$ and $p_i = p|_{\Omega_i}$

$$
L_i^* p_i = -k_i \Delta p_i, \quad i = 1, \dots, n,
$$

and for the operator B^*

$$
B^*p = p \ + \ \text{interface conditions, comp. (4.79).}
$$

As defined in the previous section S^* is the solution operator of the adjoint problem, that means

$$
p = S^*(y - y_d).
$$

and P the affine operator that maps an control u to the corresponding adjoint state p, i.e.,

$$
p = S^*(Su - y_d) = Pu.
$$

93

Remark 5.15. In the setting of Section 5.1 we have $Z = U = L^2(\Omega)$. Existence and uniqueness of a solution follow from Theorem 5.4. Lemma 5.6 is valid where the projection operator $\Pi_{U^{\mathrm{ad}}}$ is given as

$$(\Pi_{U^{\mathrm{ad}}} f)(x) = \max(u_a, \min(u_b, f(x))$$

for a continuous function f.

In the following subsections we treat each of the scalar elliptic state equations mentioned above. For the state equations (4.79) and (4.43) with (4.44) and (4.46) we will check the assumptions of Theorems 5.7 and 5.14 to derive L^2-error estimates. For the state equation (4.1) L^2-error estimates were already proved in [15, 71, 95]. The proofs of our assumptions here would be very similar to the proofs given in [15] so that we do not repeat them. Instead, we derive in the following subsection L^∞-error estimates for that case.

5.2.1 Polygonal domain

In this subsection we consider the case of a linear-quadratic optimal control problem with elliptic state equation in a polygonal domain (comp. (4.1)). As in Section 4.1 we can restrict our considerations to the case of one corner with interior angle ω located at the origin. The main focus in this subsection is on L^∞-error estimates.

5.2.1.1 Regularity

Lemma 5.16. *For the solution $\bar{u}$ of the optimal control problem (5.33) with state equation (4.1) and the associated state $\bar{y} = S\bar{u}$ and adjoint state $\bar{p} = P\bar{u}$ one has $\bar{u} \in C^{0,\sigma}(\bar{\Omega})$, $\bar{y} \in C^{0,\sigma}(\bar{\Omega}) \cap V_\beta^{2,p}(\Omega)$ and $\bar{p} \in C^{0,\sigma}(\bar{\Omega}) \cap V_\beta^{2,p}(\Omega)$ for some $\sigma \in (0,1]$, for all $p \in (1,\infty)$ and $\beta > 2 - \lambda - 2/p$. The a priori estimates*

$$\|\bar{y}\|_{C^{0,\sigma}(\bar{\Omega})} + \|\bar{y}\|_{V_\beta^{2,p}(\Omega)} \leq c\|\bar{u}\|_{L^\infty(\Omega)} \leq c\|\bar{u}\|_{C^{0,\sigma}(\bar{\Omega})}, \tag{5.34}$$

$$\|\bar{p}\|_{C^{0,\sigma}(\bar{\Omega})} + \|\bar{p}\|_{V_\beta^{2,p}(\Omega)} \leq c\left(\|\bar{u}\|_{C^{0,\sigma}(\bar{\Omega})} + \|\bar{y}_d\|_{C^{0,\sigma}(\bar{\Omega})}\right) \tag{5.35}$$

hold.

Proof. This proof reuses ideas of [15, Remark 2]. From the definition of the optimal control problem (5.33) one knows that $\bar{u} \in L^\infty(\Omega)$. This means Lemma 4.1 yields directly $\bar{y} \in V_\beta^{2,p}(\Omega)$ and

$$\|\bar{y}\|_{V_\beta^{2,p}(\Omega)} \leq c\|\bar{u}\|_{V_\beta^{0,p}(\Omega)} \leq c\|\bar{u}\|_{L^\infty(\Omega)} \tag{5.36}$$

for $p \in (1,\infty)$ and $\beta > 2 - \lambda - 2/p$. The last estimate is valid due to the embedding $L^\infty(\Omega) \hookrightarrow V_\beta^{0,p}(\Omega)$, which can be concluded from the computation

$$\|u\|_{V_\beta^{0,p}(\Omega)}^p = \int_\Omega r^{p\beta}|u|^p \, dx \leq \|u\|_{L^\infty(\Omega)}^p \int_\Omega r^{p\beta} \, dx.$$

The last integral is finite since $p\beta > (2 - \lambda)p - 2 > -2$.

Let us now prove the Hölder continuity of $\bar{y}$. Choose $p' \in (1, 2/(2-\lambda))$. Then one has $2 - \lambda - 2/p' < 0$ and therefore one can choose $\beta = 0$ in Lemma 4.1. Since $V_0^{2,p'}(\Omega) \hookrightarrow W^{2,p'}(\Omega)$ it follows $y \in W^{2,p'}(\Omega)$. For $0 < \sigma < 2 - 2/p'$ this space is embedded in $C^{0,\sigma}(\bar{\Omega})$, see Theorem 2.13. This fact allows to conclude $\bar{y} \in C^{0,\sigma}(\bar{\Omega})$. This assertion holds also for $p \geq 2/(2 - \lambda)$ because the right-hand side of the state equation is, of course, allowed to be smoother without affecting the regularity of $\bar{y}$ negatively. With the help of the a priori estimates in Lemma 4.1 one can finally conclude

$$\|\bar{y}\|_{C^{0,\sigma}(\bar{\Omega})} \leq c\|\bar{y}\|_{W^{2,p}(\Omega)} \leq c\|\bar{y}\|_{V_0^{2,p}(\Omega)} \leq c\|\bar{u}\|_{V_0^{0,p}(\Omega)} = c\|\bar{u}\|_{L^p(\Omega)} \leq c\|\bar{u}\|_{L^\infty(\Omega)}. \quad (5.37)$$

Together with the embedding $C^{0,\sigma}(\bar{\Omega}) \hookrightarrow L^\infty(\Omega)$ the estimates (5.36) and (5.37) yield the assertion (5.34). Since $y_d \in C^{0,\sigma}(\bar{\Omega})$ and therefore $y - y_d \in C^{0,\sigma}(\bar{\Omega})$ the same argumentation as above yields the stated regularity and a priori estimate (5.35) for $\bar{p}$. From the projection formula (5.6) one can finally conclude $\bar{u} \in C^{0,\sigma}(\bar{\Omega})$. $\qquad\square$

5.2.1.2 Approximation error estimate in $L^\infty(\Omega)$

We prove estimates of the pointwise error in all three variables on polygonal domains that are discretized by isotropic graded triangular meshes. The state and adjoint state equation are discretized by a finite element scheme as introduced in Subsection 4.1.2. This means, the approximate state $y_h = S_h u$ is the unique solution of

$$a_s(y_h, v_h) = (u, v_h)_{L^2(\Omega)} \quad \forall v_h \in V_{0h}$$

with bilinear form a_s defined in (4.3) and

$$V_{0h} = \left\{ v \in C(\bar{\Omega}) : v|_T \in \mathcal{P}_1 \; \forall T \in T_h \text{ and } v_h = 0 \text{ on } \partial\Omega \right\}.$$

Similarly, the approximated adjoint state $p_h = S_h^*(y - y_d)$ is the unique solution of

$$a_s(v_h, p_h) = (y_h - y_d, v_h)_{L^2(\Omega)} \quad \forall v_h \in V_h.$$

In the case of $\mu > \frac{1}{2}$ the following lemma about the boundedness of S_h and S_h^* was already proved in [15]. Since in our setting here $\mu < \lambda/2$ is required (comp. section 4.1.2), we cannot fulfill this condition for $\lambda \in [1/2, 1]$. Therefore we give in the following a more involved proof of [15, Lemma 3] without the condition $\mu > \frac{1}{2}$. Notice, that we use in this subsection again the splitting of Ω into subdomains Ω_j with corresponding quantities d_j as introduced in Chapter 4 on page 39.

Lemma 5.17. *Let T_h be a graded mesh according to (2.12) with parameter $\mu < \lambda$. The norms of the discrete solution operators S_h and S_h^* are bounded,*

$$\|S_h\|_{L^2(\Omega) \to L^\infty(\Omega)} \leq c, \qquad\qquad \|S_h^*\|_{L^2(\Omega) \to L^\infty(\Omega)} \leq c,$$

$$\|S_h\|_{L^2(\Omega) \to L^2(\Omega)} \leq c, \qquad\qquad \|S_h^*\|_{L^2(\Omega) \to L^2(\Omega)} \leq c,$$

$$\|S_h\|_{L^2(\Omega) \to H_0^1(\Omega)} \leq c, \qquad\qquad \|S_h^*\|_{L^2(\Omega) \to H_0^1(\Omega)} \leq c,$$

$$\|S_h\|_{L^\infty(\Omega) \to L^\infty(\Omega)} \leq c, \qquad\qquad \|S_h^*\|_{L^\infty(\Omega) \to L^\infty(\Omega)} \leq c,$$

where c is independent of h.

For the proof of this lemma we use estimates of norms of a regularized Green function. Before we formulate the proof of Lemma 5.17 we define this function and give some estimates on that in the forthcoming lemma and corollary. The ideas of this technique are taken from [108, Subsection 3.4.1]. We introduce the regularized Dirac function for an arbitrary but fixed element T_* as

$$\delta^h := \begin{cases} |T_*|^{-1}\operatorname{sgn}(e) & \text{in } T_*, \\ 0 & \text{elsewhere,} \end{cases} \tag{5.38}$$

where we abbreviated the finite element error by e,

$$e := y - y_h.$$

The regularized Green function g^h is defined as a solution of

$$a_s\left(\varphi, g^h\right) = \left(\delta^h, \varphi\right) \quad \forall \varphi \in V, \tag{5.39}$$

and its discrete counterpart g_h^h by

$$a_s\left(\varphi_h, g_h^h\right) = \left(\delta^h, \varphi_h\right) \quad \forall \varphi_h \in V_{0h}.$$

Lemma 5.18. *The norms of the regularized Green function can be estimated by*

$$\left\|g^h\right\|_{L^\infty(\Omega)} \le c\left|\ln h\right| \tag{5.40}$$

$$\left\|g^h\right\|_{H^1(\Omega)} \le c\left|\ln h\right|^{1/2} \tag{5.41}$$

$$\left\|g^h\right\|_{V_\beta^{2,2}(\Omega)} \le ch^{-1}, \tag{5.42}$$

where $\beta := 1 - \mu > 1 - \lambda$ is the weight corresponding to the regularity in $V_\beta^{2,2}(\Omega)$, and the grading parameter satisfies $\mu < \lambda$.

Proof. Let $g \in W_0^{1,q}(\Omega)$, $1 \le q < 2$, be the Green function with respect to an arbitrary point $x_+ \in \Omega$,

$$a_s(\varphi, g) = \varphi(x_+) \quad \forall \varphi \in W_0^{1,q'}(\Omega) \tag{5.43}$$

where $q' > 2$ satisfies $1/q + 1/q' = 1$. The Green function satisfies the inequality

$$|g(x)| \le c\left(\left|\ln\left|x - x_+\right|\right| + 1\right) \quad \forall x \in \Omega.$$

In [55] it is proven that this estimate is valid on Lipschitz domains and that the constant c is independent of x_+. Let us fix $q' = 2 + \varepsilon$. Since $\delta_h \in L^{q'}(\Omega)$ it follows from Lemma 4.1 that $g^h \in V_\beta^{2,q'}(\Omega)$ for $\beta > 2 - \lambda - 2/q'$. Since $\lambda > 1/2$ one can find $\beta < 1$ satisfying this condition and it follows from the embedding $V_\beta^{2,q'}(\Omega) \hookrightarrow V_0^{2-\beta,q'}(\Omega) \hookrightarrow W^{1,q'}(\Omega)$, see [113, Lemma 1.2], that $g^h \in W_0^{1,q'}(\Omega)$. Using (5.43), (5.39), and (5.38) we get

$$\left|g^h\left(x_+\right)\right| = \left|a_s\left(g^h, g\right)\right| = \left|\left(\delta^h, g\right)\right| \le |T_*|^{-1}\int_{T_*} |g| \, dx.$$

In the case $\text{dist}(x_+, T_*) > h_{T_*}$ we have $|x - x_+| \geq h_{T_*}$ and estimate (5.40) is obtained via

$$|T_*|^{-1} \int_{T_*} |g| \, dx \leq \max_{x \in T_*} |g(x)| \leq c \max_{x \in T_*} \left(|\ln |x - x_+|| + 1 \right) \leq c \, |\ln h_{T_*}| \leq c \, |\ln h|,$$

since $h_{T_*} \geq ch^{1/\mu}$. In the case $\text{dist}(x_+, T_*) \leq h_{T_*}$ we calculate the integral by using polar coordinates centered in x_+,

$$|T_*|^{-1} \int_{T_*} |g| \, dx \leq c|T_*|^{-1} \int_0^{2h_{T_*}} (-\ln r) r \, dr = c|T_*|^{-1} h_{T_*}^2 (c - \ln h_{T_*}) \leq c \, |\ln h|$$

as above.

For the proof of (5.41) we use the coercivity of the bilinear form and the definitions (5.39) of g^h and (5.38) of δ_h,

$$c \left\| g^h \right\|_{H^1(\Omega)}^2 \leq a_s \left(g^h, g^h \right) = \left(\delta^h, g^h \right) \leq \left\| g^h \right\|_{L^\infty(\Omega)} \left\| \delta_h \right\|_{L^1(\Omega)} \leq \left\| g^h \right\|_{L^\infty(\Omega)}.$$

With inequality (5.40) we conclude estimate (5.41).

The a priori estimate for the solution of the elliptic partial differential equation in Lemma 4.1 and the definition (5.38) of δ_h give

$$\left\| g^h \right\|_{V_\beta^{2,2}(\Omega)} \leq c \left\| r^\beta \delta^h \right\|_{L^2(\Omega)} \leq c|T_*|^{-1} \left\| r^\beta \right\|_{L^2(T_*)}.$$

With $r \leq d_J$, we can continue by

$$|T_*|^{-1} \left\| r^\beta \right\|_{L^2(T_*)} < c|T_*|^{-1/2} d_J^\beta = ch^{-1}$$

since $|T_*|^{1/2} = ch_{T_*} = chd_J^{1-\mu} = chd_J^\beta$ for $J < I$. In the other case, $J = I$, we calculate the L^2-norm and obtain

$$|T_*|^{-1} \left\| r^\beta \right\|_{L^2(T_*)} \leq c|T_*|^{-1} h_{T_*}^{\beta+1} \leq ch_{T_*}^{\beta-1} = ch^{-1}$$

since $h_{T_*} = ch^{1/\mu} = ch^{1/(1-\beta)}$. Thus inequality (5.42) is proved. $\square$

Corollary 5.19. *On meshes with grading parameter $\mu = 1 - \beta < \lambda$ the error estimates*

$$\left\| g^h - g_h^h \right\|_{H^1(\Omega)} \leq c$$
$$\left\| g^h - g_h^h \right\|_{L^2(\Omega)} \leq ch$$

hold.

Proof. Since the meshes are optimally graded, one has from Theorem 4.3

$$\left\| g^h - g_h^h \right\|_{H^1(\Omega)} \leq ch \left\| r^\beta \nabla^2 g^h \right\|_{L^2(\Omega)},$$
$$\left\| g^h - g_h^h \right\|_{L^2(\Omega)} \leq ch^2 \left\| r^\beta \nabla^2 g^h \right\|_{L^2(\Omega)}.$$

With (5.42) we get the assertion. $\square$

Now we are able to prove Lemma 5.17.

Proof of Lemma 5.17. First we prove the boundedness of $\|S_h\|_{L^2(\Omega)\to L^\infty(\Omega)}$. To this end, we consider a function $f \in L^2(\Omega)$ and an arbitrary but fixed finite element T_*. Then the three inequalities

$$\|y_h\|_{L^\infty(T_*)} \leq c|T_*|^{-1}\|y_h\|_{L^1(T_*)},$$
$$\|y_h\|_{L^1(T_*)} \leq \|y - y_h\|_{L^1(T_*)} + \|y\|_{L^1(T_*)},$$
$$\|y\|_{L^1(T_*)} \leq |T_*|\|y\|_{L^\infty(T_*)},$$

yield the estimate

$$\|y_h\|_{L^\infty(T_*)} \leq c|T_*|^{-1}\|y - y_h\|_{L^1(T_*)} + c\|y\|_{L^\infty(T_*)}. \tag{5.44}$$

By the definition of δ^h and g^h we get for the first term on the right-hand side of this inequality the equation

$$|T_*|^{-1}\|e\|_{L^1(T_*)} = (\delta_h, e) = a_s\left(e, g^h\right).$$

Using the Galerkin orthogonality and the fact that $e - I_h e = y - I_h y$ yields

$$|T_*|^{-1}\|e\|_{L^1(T_*)} = a_s\left(e, g^h - g_h^h\right) = a_s\left(e - I_h e, g^h - g_h^h\right) = a_s\left(y - I_h y, g^h - g_h^h\right).$$

With the Cauchy–Schwarz inequality we can continue

$$|T_*|^{-1}\|e\|_{L^1(T_*)} \leq c\|y - I_h y\|_{H^1(\Omega)}\left\|g^h - g_h^h\right\|_{H^1(\Omega)}. \tag{5.45}$$

From finite element theory one knows that

$$\|y - I_h y\|_{H^1(\Omega)} \leq ch^\kappa \|f\|_{L^2(\Omega)} \tag{5.46}$$

with $\kappa = \min\{\frac{\lambda}{\mu}, 1\}$, i.e., $\kappa = 1$ for $\mu < \lambda$, see [16]. Consequently, one can conclude from (5.45) together with (5.46) and Corollary 5.19

$$|T_*|^{-1}\|e\|_{L^1(T_*)} \leq ch\|f\|_{L^2(\Omega)}.$$

This shows together with (5.44) and $y_h = S_h f$

$$\|S_h f\|_{L^\infty(T_*)} \leq c\|f\|_{L^2(\Omega)},$$

where we have used the boundedness of S as operator from $L^2(\Omega)$ to $L^\infty(\Omega)$ (comp. [15, Remark 2]) to estimate the second term of the right-hand side of inequality (5.44). The boundedness of $\|S_h\|_{L^2(\Omega)\to L^2(\Omega)}$ and $\|S_h\|_{L^\infty(\Omega)\to L^\infty(\Omega)}$ follows then by the embedding $L^\infty(\Omega) \hookrightarrow L^2(\Omega)$. The boundedness of $\|S_h\|_{L^2(\Omega)\to H_0^1(\Omega)}$ comes from the theory of weak solutions. The estimates for S_h^* follow by analogy. $\square$

Variational discrete approach   First we consider the variational discrete approach that has been introduced in Subsection 5.1.2.1. In the case of convex domains and quasi-uniform triangulations, it is shown in [71] that a finite element discretization of S with piecewise linear and globally continuous functions yields an approximation rate of 2 in the L^2-norm. These results extend to nonconvex domains with graded meshes (see [15, Remark 5]). Therefore the following theorem is valid.

Theorem 5.20. *Let $\bar{u}$ be the solution of the optimal control problem (5.33) with the state equation (4.1) and $\bar{u}_h^s$ the solution of the corresponding variational discrete problem (5.8) on a mesh of type (2.12) with grading parameter $\mu < \lambda$. Then the estimate*

$$\|\bar{u} - \bar{u}_h^s\|_{L^2(\Omega)} \leq ch^2 \left(\|\bar{u}\|_{L^2(\Omega)} + \|y_d\|_{L^2(\Omega)} \right)$$

is valid.

With the L^2-error estimate at hand, one is now able to prove an L^∞-error estimate.

Theorem 5.21. *Let $\bar{u}_h^s$ be the discrete control introduced in (5.7) and $\bar{y}_h^s = S_h \bar{u}_h^s$ and $\bar{p}_h^s = P_h \bar{u}_h^s$ the associated state and adjoint state, respectively. On a family of meshes with grading parameter $\mu < \frac{\lambda}{2}$ the estimates*

$$\|\bar{u} - \bar{u}_h^s\|_{L^\infty(\Omega)} \leq ch^2 |\ln h|^{3/2} \left(\|\bar{u}\|_{C^{0,\sigma}(\bar{\Omega})} + \|y_d\|_{C^{0,\sigma}(\bar{\Omega})} \right) \tag{5.47}$$

$$\|\bar{y} - \bar{y}_h^s\|_{L^\infty(\Omega)} \leq ch^2 |\ln h|^{3/2} \left(\|\bar{u}\|_{C^{0,\sigma}(\bar{\Omega})} + \|y_d\|_{C^{0,\sigma}(\bar{\Omega})} \right) \tag{5.48}$$

$$\|\bar{p} - \bar{p}_h^s\|_{L^\infty(\Omega)} \leq ch^2 |\ln h|^{3/2} \left(\|\bar{u}\|_{C^{0,\sigma}(\bar{\Omega})} + \|y_d\|_{C^{0,\sigma}(\bar{\Omega})} \right) \tag{5.49}$$

are valid.

Proof. The ideas of the proof are similar to that given in [71] for quasi-uniform meshes and $W^{2,\infty}(\Omega)$-regular solutions of the underlying boundary value problem. First, we prove assertion (5.49). One can conclude

$$\begin{aligned}
\|\bar{p} - \bar{p}_h^s\|_{L^\infty(\Omega)} &= \|S^*(S\bar{u} - y_d) - S_h^*(S_h \bar{u}_h^s - y_d)\|_{L^\infty(\Omega)} \\
&\leq \|(S^* - S_h^*)S\bar{u}\|_{L^\infty(\Omega)} + \|(S^* - S_h^*)y_d\|_{L^\infty(\Omega)} \\
&\quad + \|S_h^* S\bar{u} - S_h^* S_h \bar{u}\|_{L^\infty(\Omega)} + \|S_h^* S_h \bar{u} - S_h^* S_h \bar{u}_h^s\|_{L^\infty(\Omega)}. \tag{5.50}
\end{aligned}$$

We estimate each of the four terms separately. By Theorem 4.4 it follows

$$\|(S^* - S_h^*)S\bar{u}\|_{L^\infty(\Omega)} \leq ch^2 |\ln h|^{3/2} \|S\bar{u}\|_{C^{0,\sigma}(\bar{\Omega})} \leq ch^2 |\ln h|^{3/2} \|\bar{u}\|_{C^{0,\sigma}(\bar{\Omega})} \tag{5.51}$$

since $\|S\|_{C^{0,\sigma}(\bar{\Omega}) \to C^{0,\sigma}(\bar{\Omega})} \leq c$. For the second term one can conclude from the same theorem

$$\|(S^* - S_h^*)Sy_d\|_{L^\infty(\Omega)} \leq ch^2 |\ln h|^{3/2} \|y_d\|_{C^{0,\sigma}(\bar{\Omega})}. \tag{5.52}$$

With the discrete Sobolev inequality (4.9) one has

$$\|S_h^* S\bar{u} - S_h^* S_h \bar{u}\|_{L^\infty(\Omega)} \le c\,|\ln h|^{1/2}\,\|S_h^* S\bar{u} - S_h^* S_h \bar{u}\|_{H^1(\Omega)}$$
$$\le c\,|\ln h|^{1/2}\,\|S_h^*\|_{L^2(\Omega)\to H^1(\Omega)}\|S\bar{u} - S_h\bar{u}\|_{L^2(\Omega)}$$
$$\le c\,|\ln h|^{1/2}\,h^2\|\bar{u}\|_{L^2(\Omega)}, \tag{5.53}$$

where we have used Theorem 4.3 in the last step. Utilizing again inequality (4.9), Lemma 5.17, and Theorem 5.20 it follows for the fourth term

$$\|S_h^* S_h \bar{u} - S_h^* S_h \bar{u}_h^s\|_{L^\infty(\Omega)} \le c\,|\ln h|^{1/2}\,\|S_h^* S_h \bar{u} - S_h^* S_h \bar{u}_h^s\|_{H^1(\Omega)}$$
$$\le c\,|\ln h|^{1/2}\,\|S_h \bar{u} - S_h \bar{u}_h^s\|_{L^2(\Omega)}$$
$$\le c\,|\ln h|^{1/2}\,\|\bar{u} - \bar{u}_h^s\|_{L^2(\Omega)}$$
$$\le c\,|\ln h|^{1/2}\,h^2\left(\|\bar{u}\|_{L^2(\Omega)} + \|y_d\|_{L^2(\Omega)}\right). \tag{5.54}$$

The estimate (5.50) yields together with (5.51)–(5.54) the assertion (5.49). Since the condition (5.7) is equivalent to the expression $\bar{u}_h^s = \Pi_{U^{\mathrm{ad}}}(-\frac{1}{\nu}\bar{p}_h^s)$ (comp. (5.6)) one can conclude

$$\|\bar{u} - \bar{u}_h^s\|_{L^\infty(\Omega)} \le \frac{1}{\nu}\|\bar{p} - \bar{p}_h^s\|_{L^\infty(\Omega)}$$

and inequality (5.47) follows directly from (5.49). To show inequality (5.48) we conclude

$$\|y - \bar{y}_h^s\|_{L^\infty(\Omega)} \le \|S\bar{u} - S_h\bar{u}\|_{L^\infty(\Omega)} + \|S_h\bar{u} - S_h\bar{u}_h^s\|_{L^\infty(\Omega)}$$
$$\le c h^2\,|\ln h|^{3/2}\left(\|\bar{u}\|_{C^{0,\sigma}(\bar{\Omega})} + \|y_d\|_{C^{0,\sigma}(\bar{\Omega})}\right),$$

where we have used Theorem 4.4, Lemma 5.17, and inequality (5.47) in the last step. $\square$

Postprocessing approach The aim of this paragraph is to prove error estimates of the same quality as in Theorem 5.21 but now for the postprocessing approach introduced in Subsection 5.1.2.2.

The optimal control $\bar{u}$ is obtained by the projection formula (5.6). This formula generates kinks in the optimal control. However, we can classify the triangles $T \in T_h$ in two sets K_1 and K_2,

$$K_1 := \bigcup_{T\in T_h:\,\bar{u}\notin V^{2,2}_{2-2\mu}(T)} T, \qquad K_2 := \bigcup_{T\in T_h:\,\bar{u}\in V^{2,2}_{2-2\mu}(T)} T.$$

This means, K_1 contains the elements that have a nonempty intersection with both the active and the inactive set. Clearly, the number of triangles in K_1 grows for decreasing h. Nevertheless, the assumption

$$\mathrm{meas}\,K_1 \le ch \tag{5.55}$$

is fulfilled in many practical cases. Next, we recall a supercloseness result (comp. Theorem 5.13) and a auxiliary result (comp. Assumption PP4) from [15].

Theorem 5.22. *Assume that Assumption* (5.55) *holds. Let $\bar{u}_h$ be the solution of* (5.12) *with state equation* (4.4) *on a family of meshes with grading parameter $\mu < \lambda$. Then the estimate*

$$\|\bar{u}_h - R_h\bar{u}\|_{L^2(\Omega)} \leq ch^2 \left(\|\bar{u}\|_{L^\infty(\Omega)} + \|y_d\|_{L^\infty(\Omega)}\right)$$

holds true.

Proof. This theorem is proved in [15] under the assumption $\mu > \frac{1}{2}$, which was used in the proof of the boundedness of S_h only. The boundedness of S_h also in case of $\mu \leq 1/2$ is guaranteed by Lemma 5.17. $\qquad\square$

Lemma 5.23. *On a mesh with grading parameter $\mu < \lambda$ the estimate*

$$(v_h, \bar{u} - R_h\bar{u})_{L^2(\Omega)} \leq ch^2 \left(\|v_h\|_{L^\infty(\Omega)} + \|v_h\|_{H_0^1(\Omega)}\right) \left(\|\bar{u}\|_{L^\infty(\Omega)} + \|\bar{y}_d\|_{L^\infty(\Omega)}\right)$$

holds for all $v_h \in V_h$, provided that Assumption (5.55) *is fulfilled.*

Proof. This lemma is proved in [15]. Notice that in that proof the condition $\mu \geq 1/2$ was not necessary. $\qquad\square$

Next, we will apply the error estimates of subsection 4.1.2 to obtain L^∞-error estimates for the optimal control problem. Before, we will derive an auxiliary result. To this end, we define another regularized Dirac function δ_ξ^h for a fixed point $\xi \in T_*$ with

(P1) $(\delta_\xi^h, v_h) = v_h(\xi) \qquad \forall v_h \in V_{0h}$,

(P2) $\operatorname{supp}\delta_\xi^h \subset \bar{T}_*$,

(P3) $\delta_\xi^h \in \mathcal{P}_1(T_*)$,

(P4) $\|\delta_\xi^h\|_{L^2(T_*)} = O(h_{T_*}^{-1})$,

(P5) $\|\delta_\xi^h\|_{L^\infty(\Omega)} \leq c|T_*|^{-1}$.

An example for a function with these properties is given in [121]. The regularized Green function z^h is defined as the solution of

$$a_s\left(v, z^h\right) = \left(\delta_\xi^h, v\right) \qquad \forall v \in V. \tag{5.56}$$

Moreover, we denote by z_h^h its discrete counterpart,

$$a_s\left(v_h, z_h^h\right) = \left(\delta_\xi^h, v_h\right) \qquad \forall v_h \in V_{0h}.$$

Subsequently, we need estimates of norms of the regularized Green function.

Lemma 5.24. *The norms of the regularized Green function can be estimated by*

$$\left\|z^h\right\|_{L^\infty(\Omega)} \leq c\,|\ln h| \tag{5.57}$$

$$\left\|z^h\right\|_{H^1(\Omega)} \leq c\,|\ln h|^{1/2} \tag{5.58}$$

$$\left\|z^h\right\|_{V_{1-\mu}^{2,2}(\Omega)} \leq ch^{-1}, \tag{5.59}$$

with grading parameter $\mu < \lambda$ in (2.12).

Proof. The proof of this lemma is very similar to that of Lemma 5.18. For the sake of completeness we sketch it here. Let $g \in W_0^{1,q}(\Omega)$, $1 \leq q < 2$, be the Green function with respect to an arbitrary point $x_+ \in \Omega$,

$$a_s(g,v) = v(x_+) \qquad \forall v \in W_0^{1,q'}(\Omega) \tag{5.60}$$

where $q' > 2$ and $1/q + 1/q' = 1$. According to [55] there is a constant c independent of x_+, such that

$$|g(x)| \leq c(|\ln|x - x_+|| + 1).$$

With the same argumentation as for g^h in Lemma 5.18 one can conclude $z^h \in W_0^{1,q'}(\Omega)$, $q' = 2 + \varepsilon$. Using (5.60), (5.56), the Hölder inequality and property (P4) we get

$$\left|z^h(x_+)\right| = \left|a_s\left(g, z^h\right)\right| = (\delta_\xi^h, g) \leq \left\|\delta_\xi^h\right\|_{L^2(T_*)} \|g\|_{L^2(T_*)} \leq c h_{T_*}^{-1} \|g\|_{L^2(T_*)}. \tag{5.61}$$

We estimate the L^2-norm of g using polar coordinates centered in x_+,

$$h_{T_*}^{-1}\|g\|_{L^2(T_*)} \leq c h_{T_*}^{-1} \left(\int_0^{h_{T_*}} (\ln r)^2 r dr\right)^{1/2}$$

$$\leq c h_{T_*}^{-1}\left(h_{T_*}|\ln h_{T_*}| + h_{T_*}|\ln h_{T_*}|^{1/2} + h_{T_*}\right)$$

$$\leq c|\ln h|.$$

This yields from (5.61) the estimate (5.57).

For the proof of (5.58) we use the coercivity of the bilinear form, the definition (5.56), and property (P4) of $\delta_h(a)$,

$$\left\|z^h\right\|_{H^1(\Omega)}^2 \leq c \cdot a\left(z^h, z^h\right) = c\left(\delta^h, z^h\right) \leq c\left\|z^h\right\|_{L^\infty(\Omega)} \|\delta_h\|_{L^1(\Omega)}$$

$$\leq c|T_*|^{1/2}\|\delta_h\|_{L^2(T_*)}\left\|z^h\right\|_{L^\infty(\Omega)} \leq c\left\|z^h\right\|_{L^\infty(\Omega)}.$$

With inequality (5.57) we conclude estimate (5.58).

The a priori estimate for the solution of the elliptic partial differential equation stated in Lemma 4.1 and the property (P5) of $\delta_h(\xi)$ give

$$\left\|z^h\right\|_{V_\beta^{2,2}(\Omega)} \leq c\left\|r^\beta \delta^h\right\|_{L^2(\Omega)} \leq c|T_*|^{-1}\left\|r^\beta\right\|_{L^2(T_*)}.$$

In the case $J < I$ it is $r \leq d_J$ and we can continue by

$$|T_*|^{-1}\left\|r^\beta\right\|_{L^2(T_*)} \leq c|T_*|^{-1/2}d_J^\beta = ch^{-1}$$

since $|T_*|^{1/2} = ch_{T_*} = chd_J^{1-\mu} = chd_J^\beta$. In the other case, $J = I$, we calculate the L^2-norm and obtain

$$|T_*|^{-1}\left\|r^\beta\right\|_{L^2(T_*)} \leq c|T_*|^{-1}h_{T_*}^{\beta+1} \leq ch_{T_*}^{\beta-1} = ch^{-1}$$

since $h_{T_*} = ch^{1/\mu} = ch^{1/(1-\beta)}$. Thus, (5.59) is proved. $\qquad\square$

Lemma 5.25. *The estimate*

$$\left\|z_h^h\right\|_{L^\infty(\Omega)} + \left\|z_h^h\right\|_{H_0^1(\Omega)} \leq c\left|\ln h\right|$$

holds on a finite element mesh of type (2.12) with $\mu < \lambda$.

Proof. By the triangle inequality one can conclude

$$\left\|z_h^h\right\|_{L^\infty(\Omega)} + \left\|z_h^h\right\|_{H_0^1(\Omega)}$$
$$\leq \left\|z^h\right\|_{L^\infty(\Omega)} + \left\|z^h - z_h^h\right\|_{L^\infty(\Omega)} + \left\|z_h\right\|_{H_0^1(\Omega)} + \left\|z^h - z_h^h\right\|_{H_0^1(\Omega)}. \tag{5.62}$$

Since the meshes are optimally graded, we have from [16]

$$\left\|z^h - z_h^h\right\|_{H^1(\Omega)} \leq ch\left|z^h\right|_{V_\beta^{2,2}(\Omega)}$$

and with (5.59)

$$\left\|z^h - z_h^h\right\|_{H^1(\Omega)} \leq c. \tag{5.63}$$

Furthermore, we have

$$\left\|z^h - z_h^h\right\|_{L^\infty(\Omega)} \leq ch\left|z^h\right|_{V_\beta^{2,2}(\Omega)}$$

and with (5.59)

$$\left\|z^h - z_h^h\right\|_{L^\infty(\Omega)} \leq c. \tag{5.64}$$

Now the assertion follows from (5.62) with (5.57), (5.58), (5.63), and (5.64). □

The following lemma is an analogous estimate to (5.17) in $L^\infty(\Omega)$.

Lemma 5.26. *The inequality*

$$\left\|S_h\bar{u} - S_h R_h\bar{u}\right\|_{L^\infty(\Omega)} \leq ch^2\left|\ln h\right|\left(\left\|\bar{u}\right\|_{L^\infty(\Omega)} + \left\|\bar{y}_d\right\|_{L^\infty(\Omega)}\right)$$

is satisfied provided that Assumption (5.55) is fulfilled.

Proof. Let $\xi \in \Omega$ be an arbitrary but fixed point. Using the definitions above, we find

$$|S_h\bar{u}(\xi) - S_h R_h\bar{u}(\xi)| = \left|\left(\delta_\xi^h, S_h\bar{u} - S_h R_h\bar{u}\right)\right|$$
$$= \left|a\left(S_h\bar{u} - S_h R_h\bar{u}, z_h^h\right)\right|$$
$$= \left|\left(z_h^h, \bar{u} - R_h\bar{u}\right)\right|.$$

Now, we can apply Lemma 5.23 and obtain

$$|S_h\bar{u}(\xi) - S_h R_h\bar{u}(\xi)| \leq ch^2\left(\left\|z_h^h\right\|_{L^\infty(\Omega)} + \left\|z_h^h\right\|_{H_0^1(\Omega)}\right)\left(\left\|\bar{u}\right\|_{L^\infty(\Omega)} + \left\|\bar{y}_d\right\|_{L^\infty(\Omega)}\right). \tag{5.65}$$

The assertion follows from (5.65) and Lemma 5.25. □

The following L^∞-error estimates are new even in the case without corner singularities and quasi-uniform meshes; see [95].

Theorem 5.27. *Assume that Assumption (5.55) holds. Let $\bar{y}_h$ be the associated state and $\bar{p}_h$ be the associated adjoint state to the solution $\bar{u}_h$ of (5.12) with state equation (4.4) on meshes of type (2.12) with grading parameter $\mu < \lambda/2$. Further, let $\tilde{u}_h$ be the discrete control constructed in (5.14). Then the estimates*

$$\|\bar{y}_h - \bar{y}\|_{L^\infty(\Omega)} \le ch^2 \left|\ln h\right|^{3/2} \left(\|\bar{u}\|_{C^{0,\sigma}(\bar{\Omega})} + \|y_d\|_{C^{0,\sigma}(\bar{\Omega})} \right) \tag{5.66}$$

$$\|\bar{p}_h - \bar{p}\|_{L^\infty(\Omega)} \le ch^2 \left|\ln h\right|^{3/2} \left(\|\bar{u}\|_{C^{0,\sigma}(\bar{\Omega})} + \|y_d\|_{C^{0,\sigma}(\bar{\Omega})} \right) \tag{5.67}$$

$$\|\bar{u} - \tilde{u}_h\|_{L^\infty(\Omega)} \le ch^2 \left|\ln h\right|^{3/2} \left(\|\bar{u}\|_{C^{0,\sigma}(\bar{\Omega})} + \|y_d\|_{C^{0,\sigma}(\bar{\Omega})} \right) \tag{5.68}$$

are valid.

Proof. We start with

$$\|\bar{y} - \bar{y}_h\|_{L^\infty(\Omega)} = \|S\bar{u} - S_h\bar{u}_h\|_{L^\infty(\Omega)}$$
$$\le \|S\bar{u} - S_h\bar{u}\|_{L^\infty(\Omega)} + \|S_h\bar{u} - S_hR_h\bar{u}\|_{L^\infty(\Omega)} + \|S_hR_h\bar{u} - S_h\bar{u}_h\|_{L^\infty(\Omega)}.$$

The first term was estimated in Theorem 4.4. Lemma 5.26 delivers an inequality for the second term. Theorem 5.22 implies together with $\|S_h\|_{L^2(\Omega) \to L^\infty(\Omega)} \le c$, see Lemma 5.17, the estimate of the third term. Consequently, we find with the embedding $C^{0,\sigma}(\bar{\Omega}) \hookrightarrow L^\infty(\Omega)$

$$\|\bar{y} - \bar{y}_h\|_{L^\infty(\Omega)} \le ch^2 \left|\ln h\right|^{3/2} \left(\|\bar{u}\|_{C^{0,\sigma}(\bar{\Omega})} + \|y_d\|_{L^\infty(\Omega)} \right)$$
$$\le ch^2 \left|\ln h\right|^{3/2} \left(\|\bar{u}\|_{C^{0,\sigma}(\bar{\Omega})} + \|y_d\|_{C^{0,\sigma}(\bar{\Omega})} \right),$$

i.e., estimate (5.66). The second inequality can be obtained similarly,

$$\|\bar{p} - \bar{p}_h\|_{L^\infty(\Omega)} = \|S^*(\bar{y} - y_d) - S_h^*(\bar{y}_h - y_d)\|_{L^\infty(\Omega)}$$
$$\le \|S^*(\bar{y} - y_d) - S_h^*(\bar{y} - y_d)\|_{L^\infty(\Omega)} + \|S_h^*(\bar{y} - \bar{y}_h)\|_{L^\infty(\Omega)}$$
$$\le ch^2 \left|\ln h\right|^{3/2} \|\bar{y} - y_d\|_{C^{0,\sigma}(\bar{\Omega})} + ch^2 \left|\ln h\right|^{3/2} \left(\|\bar{u}\|_{C^{0,\sigma}(\bar{\Omega})} + \|y_d\|_{C^{0,\sigma}(\bar{\Omega})} \right)$$
$$\le ch^2 \left|\ln h\right|^{3/2} \left(\|\bar{u}\|_{C^{0,\sigma}(\bar{\Omega})} + \|y_d\|_{C^{0,\sigma}(\bar{\Omega})} \right),$$

by means of Theorem 4.4, (5.66) and Lemma 5.17. To prove the third inequality we use that the projection operator $\Pi_{U^{ad}}$ is Lipschitz continuous with constant 1 from $L^\infty(\Omega)$ to $L^\infty(\Omega)$. Therefore, we get

$$\nu\|\bar{u} - \tilde{u}_h\|_{L^\infty(\Omega)} = \nu \left\| \Pi_{U^{ad}}\left(-\frac{1}{\nu}\bar{p} \right) - \Pi_{U^{ad}}\left(-\frac{1}{\nu}\bar{p}_h \right) \right\|_{L^\infty(\Omega)}$$
$$\le \|\bar{p} - \bar{p}_h\|_{L^\infty(\Omega)}$$
$$\le ch^2 \left|\ln h\right|^{3/2} \left(\|\bar{u}\|_{C^{0,\sigma}(\bar{\Omega})} + \|y_d\|_{C^{0,\sigma}(\bar{\Omega})} \right),$$

where we used (5.68) in the last step. The superconvergence result is proved. □

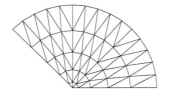 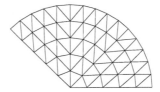

Figure 5.1: Convex domain with a graded mesh with $\mu = 0.6$ and a quasi-uniform mesh ($\mu = 1$).

5.2.1.3 Numerical tests

We illustrate our theoretical findings for the fully discrete approach by some numerical tests. Therefore we consider the optimal control problem (5.33) with the Poisson equation as state equation and the first-order optimality system

$$-\Delta \bar{y} = \bar{u} + f \ \text{ in } \Omega, \quad \bar{y} = 0 \ \text{ on } \partial\Omega,$$
$$-\Delta \bar{p} = \bar{y} - y_d \ \text{ in } \Omega, \quad p = 0 \ \text{ on } \partial\Omega,$$
$$\bar{u} = \Pi_{U^{\text{ad}}} \left(-\frac{1}{\nu} \bar{p} \right).$$

The data y_d and f are chosen such that the exact solution is given as

$$\bar{y}(r, \varphi) = \left(r^\lambda - r^\alpha \right) \sin \lambda\varphi,$$
$$\bar{p}(r, \varphi) = \nu \left(r^\lambda - r^\beta \right) \sin \lambda\varphi.$$

We set $\alpha = \beta = \frac{5}{2}$ and $\mu = 10^{-3}$. The problem is solved using the primal-dual active set strategy. For details on this we refer to [81]. To evaluate the maximum norm of the error we used not only grid points but also the nodes of a high order quadrature formula of degree 19 implemented in the program package MooNMD [75]. In the following we study the example in a convex and a nonconvex domain.

Example in a convex domain   The domain Ω is defined as

$$\Omega = \left\{ (r \cos \varphi, r \sin \varphi)^T : 0 < r < 1, 0 < \varphi < \frac{3}{4}\pi \right\},$$

and therefore $\lambda = \frac{4}{3}$. Table 5.1 shows the computed errors $\|\bar{u} - \tilde{u}_h\|_{L^\infty(\Omega)}$ and the estimated order of convergence (eoc) for quasi-uniform meshes and for graded meshes with $\mu = 0.6 < \frac{\lambda}{2}$; see Figure 5.1 for an illustration of these meshes. While a convergence rate of about λ can be observed for $\mu = 1$, the approximation order is slightly smaller than 2 on the graded meshes. So mesh grading improves the convergence rate for the L^∞-error also in the case of a corner with an interior angle between $\frac{\pi}{2}$ and π.

Table 5.1: L^∞-error of the computed control $\tilde{u}_h$ in a convex domain.

ndof	$\mu = 0.6$ value	eoc	$\mu = 1$ value	eoc
51	3.02e−02		4.63e−02	
176	1.19e−02	1.50	1.88e−02	1.45
651	4.12e−03	1.62	7.57e−03	1.39
2501	1.42e−03	1.58	3.02e−03	1.36
9801	4.22e−04	1.78	1.20e−03	1.35
38801	1.15e−04	1.89	4.79e−04	1.34
74482	6.11e−05	1.94	3.09e−04	1.34
154401	3.01e−05	1.95	1.90e−04	1.34

Table 5.2: $L^\infty(\Omega)$-error of the computed control $\tilde{u}_h$ in a nonconvex domain.

ndof	$\mu = 0.3$ value	eoc	$\mu = 0.6$ value	eoc	$\mu = 1$ value	eoc
125	1.63e−01		8.18e−02		2.19e−01	
286	7.61e−01	1.20	3.77e−02	1.23	1.34e−01	0.78
1071	2.30e−02	1.81	1.73e−02	1.18	6.67e−02	1.06
4141	7.49e−03	1.66	7.99e−03	1.14	4.15e−02	0.70
16281	1.97e−03	1.95	3.70e−03	1.13	2.60e−02	0.68
25351	1.29e−03	1.92	2.88e−03	1.12	2.24e−02	0.68
39501	8.29e−04	1.98	2.25e−03	1.12	1.93e−02	0.68
100701	3.34e−04	1.95	1.33e−03	1.12	1.41e−02	0.67

Example in a nonconvex domain As a second example we set Ω as

$$\Omega = \left\{ (r\cos\varphi, r\sin\varphi)^T : 0 < r < 1, 0 < \varphi < \frac{3}{2}\pi \right\}.$$

This means $\lambda = \frac{2}{3}$. In Table 5.2 one can find the computed errors $\|\bar{u} - \tilde{u}_h\|_{L^\infty(\Omega)}$ on different meshes with $\mu = 0.3 < \frac{\lambda}{2}$, $\mu = 0.6 < \lambda$, and $\mu = 1.0$. For meshes with grading parameter $\mu < \frac{\lambda}{2}$ one can see the predicted convergence rate slightly smaller than 2. Further one can observe that a mesh grading parameter $\mu \in \left(\frac{\lambda}{2}, \lambda\right)$ yields only a suboptimal convergence rate $\frac{\lambda}{\mu} = \frac{10}{9}$ for the L^∞-error. Notice that such a mesh grading was enough to get the optimal convergence of second order for the L^2-error (see [15]). If no mesh grading is performed ($\mu = 1$), one can observe a convergence rate of about λ. In Figure 5.2 the distribution of the L^∞-error on a quasi-uniform and on an appropriately graded mesh is shown. The mesh grading significantly reduces the error near the corner.

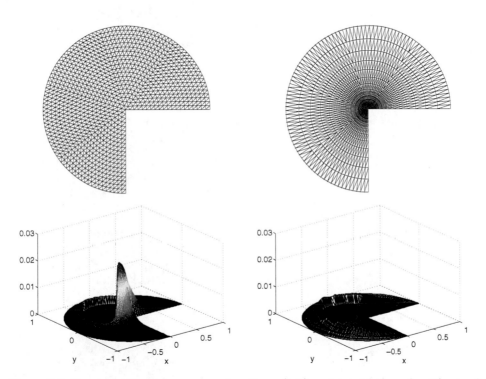

Figure 5.2: Error distribution on a quasi uniform (left) and a graded mesh with $\mu = 0.3$.

5.2.2 Prismatic domain

In this subsection we prove L^2-error estimates on anisotropic meshes for the optimal control problem (5.33) with the state equation (4.43) with Dirichlet (4.44) or Neumann boundary conditions (4.46). The results of this subsection were originally published in [18].

5.2.2.1 Regularity

Lemma 5.28. *Let $\bar{u}$ be the solution of the optimal control problem (5.33) with the state equation (4.43) with Dirichlet (4.44) or Neumann boundary conditions (4.46) and $\bar{y} = S\bar{u}$ and $\bar{p} = P\bar{u}$ the associated state and adjoint state. For both types of boundary conditions one has $\bar{u} \in C^{0,\sigma}(\bar{\Omega})$, $\bar{y} \in C^{0,\sigma}(\bar{\Omega})$ and $\bar{p} \in C^{0,\sigma}(\bar{\Omega})$ for some $\sigma \in (0, \lambda)$, $\lambda = \pi/\omega$. The a priori estimates*

$$\|\bar{y}\|_{C^{0,\sigma}(\bar{\Omega})} \le c\|\bar{u}\|_{L^\infty(\Omega)} \le c\|\bar{u}\|_{C^{0,\sigma}(\bar{\Omega})}, \tag{5.69}$$

$$\|\bar{p}\|_{C^{0,\sigma}(\bar{\Omega})} \le c\left(\|\bar{u}\|_{L^\infty(\Omega)} + \|y_d\|_{C^{0,\sigma}(\bar{\Omega})}\right) \tag{5.70}$$

are valid.

Proof. For $0 < \gamma < 2 - \frac{3}{p} - \sigma$ with p specified below the inclusion

$$V_\gamma^{2,p}(\Omega) \hookrightarrow V_0^{2-\gamma,p}(\Omega) \hookrightarrow W^{2-\gamma,p}(\Omega) \hookrightarrow C^{0,\sigma}(\bar{\Omega}) \tag{5.71}$$

is valid. For the first embedding we have used [113, Lemma 1.2]. The other inclusions follow by the Sobolev embedding theorems and the fact that $2 - \gamma - \frac{3}{p} > \sigma$. Taking the decomposition $\bar{y} = \bar{y}_r + \bar{y}_s$ into account one can conclude from Lemma 4.17 $\bar{y}_r \in W^{2,p}(\Omega)$ and $\bar{y}_s \in V_\gamma^{2,p}(\Omega)$ for $\gamma > 2 - \frac{2}{p} - \lambda$. In order to be able to find γ such that

$$2 - \frac{2}{p} - \lambda < \gamma < 2 - \frac{3}{p} - \sigma, \tag{5.72}$$

we have to choose p such that $\frac{1}{p} < \lambda - \sigma$. Since $\sigma < \lambda$, the condition $p > \frac{1}{\lambda - \sigma}$ guarantees the existence of a weight γ satisfying (5.72). With such a weight γ we can write for $p > \max\left(\frac{1}{\lambda - \sigma}, \frac{3}{2 - \sigma}\right)$ and $\sigma < \lambda$

$$\|\bar{y}\|_{C^{0,\sigma}(\bar{\Omega})} \le c\left(\|\bar{y}_s\|_{C^{0,\sigma}(\bar{\Omega})} + \|\bar{y}_r\|_{C^{0,\sigma}(\bar{\Omega})}\right)$$

$$\le c\left(\|\bar{y}_s\|_{V_\gamma^{2,p}(\Omega)} + \|\bar{y}_r\|_{W^{2,p}(\Omega)}\right)$$

$$\le c\|\bar{u}\|_{L^p(\Omega)} \le c\|\bar{u}\|_{L^\infty(\Omega)},$$

where we have used the embeddings (5.71) and $W^{2,p}(\Omega) \hookrightarrow C^{0,\sigma}(\bar{\Omega})$ for $p > \frac{3}{2-\sigma}$ as well as Lemma 4.17. This proves the first inequality in (5.69). With the same argumentation for $\bar{p}$ one gets

$$\|\bar{p}\|_{C^{0,\sigma}(\bar{\Omega})} \le c\left(\|\bar{y} - y_d\|_{L^\infty(\Omega)}\right) \le c\left(\|\bar{y}\|_{C^{0,\sigma}(\bar{\Omega})} + \|y_d\|_{C^{0,\sigma}(\bar{\Omega})}\right)$$

where we have used the triangle inequality and the embedding $C^{0,\sigma}(\bar{\Omega}) \hookrightarrow L^\infty(\Omega)$ in the last step. Together with the first inequality in (5.69) this proves the assertion (5.70). Inequality (5.70) implies together with the projection formula (5.6) that $\bar{u} \in C^{0,\sigma}(\bar{\Omega})$. The embedding $C^{0,\sigma}(\bar{\Omega}) \hookrightarrow L^\infty(\Omega)$ yields finally the second inequality of (5.69). $\quad\square$

Corollary 5.29. *Consider the optimal control problem (5.33) with state equation (4.43) and Dirichlet boundary conditions (4.44) or Neumann boundary conditions (4.46). If $\beta > 1 - \lambda$ then there holds for $i = 1, 2$ and $\sigma \in (0, \lambda)$*

$$\|r^\beta \partial_i \bar{p}\|_{L^\infty(\Omega)} \leq c \left(\|\bar{u}\|_{L^\infty(\Omega)} + \|y_d\|_{C^{0,\sigma}(\bar{\Omega})} \right), \tag{5.73}$$

$$\|\partial_3 \bar{p}\|_{L^\infty(\Omega)} \leq c \left(\|\bar{u}\|_{L^\infty(\Omega)} + \|y_d\|_{C^{0,\sigma}(\bar{\Omega})} \right). \tag{5.74}$$

Proof. From inequality (4.54) one has for some $\sigma \in (0, 1)$ the estimate

$$\|r^\beta \partial_i \bar{p}\|_{L^\infty(\Omega)} \leq c\|\bar{y} - y_d\|_{C^{0,\sigma}(\bar{\Omega})} \leq c \left(\|\bar{y}\|_{C^{0,\sigma}(\bar{\Omega})} + \|y_d\|_{C^{0,\sigma}(\bar{\Omega})} \right)$$

where we have used the triangle inequality in the last step. Now the proof of assertion (5.73) follows from inequality (5.69). If we use the estimates (4.55) and (5.69), we can conclude

$$\|\partial_3 \bar{p}\|_{L^\infty(\Omega)} \leq c\|\bar{y} - y_d\|_{C^{0,\sigma}(\bar{\Omega})} \leq c \left(\|\bar{y}\|_{C^{0,\sigma}(\bar{\Omega})} + \|y_d\|_{C^{0,\sigma}(\bar{\Omega})} \right),$$

where we have utilized the triangle inequality in the last step. The estimate (5.69) yields then the assertion (5.74). $\quad\square$

Lemma 5.30. *Let $\bar{u}$ be the solution of the optimal control problem (5.33) with the state equation (4.43) with Dirichlet boundary conditions (4.44). For the associated state $\bar{y} = S\bar{u}$ and adjoint state $\bar{p} = S\bar{u}$ one has $\bar{y} \in V_\beta^{2,p}(\Omega)$ and $\bar{p} \in V_\beta^{2,p}(\Omega)$ for $\beta > 2 - \pi/\omega - 2/p$. The a priori estimates*

$$\|\bar{y}\|_{V_\beta^{2,p}(\Omega)} \leq c\|\bar{u}\|_{L^\infty(\Omega)}, \tag{5.75}$$

$$\|\bar{p}\|_{V_\beta^{2,p}(\Omega)} \leq c \left(\|\bar{u}\|_{L^\infty(\Omega)} + \|y_d\|_{C^{0,\sigma}(\bar{\Omega})} \right) \tag{5.76}$$

hold.

Proof. Since $\bar{u} \in L^\infty(\Omega)$ and $L^\infty(\Omega) \hookrightarrow L^p(\Omega)$ for $p \in (1, \infty)$ it follows from lemma 4.15 that $\bar{y} \in V_\beta^{2,p}(\Omega)$ for $\beta > 2 - \pi/\omega - 2/p$. The estimate (5.75) follows from that Lemma with the same argumentation as in Lemma 5.16 (comp. (5.36)). From Lemma 5.28, $y_d \in C^{0,\sigma}(\bar{\Omega})$ and $C^{0,\sigma}(\bar{\Omega}) \hookrightarrow L^\infty(\Omega)$ one has $\bar{y} - y_d \in L^\infty(\Omega)$ and the same argumentation as for $\bar{y}$ yields $\bar{p} \in V_\beta^{2,p}(\Omega)$ for $\beta > 2 - \pi/\omega - 2/p$. The estimate

$$\|\bar{p}\|_{V_\beta^{2,p}(\Omega)} \leq c \left(\|\bar{y}\|_{V_\beta^{0,p}(\Omega)} + \|y_d\|_{V_\beta^{0,p}(\Omega)} \right) \leq c \left(\|\bar{y}\|_{C^{0,\sigma}(\bar{\Omega})} + \|y_d\|_{C^{0,\sigma}(\bar{\Omega})} \right)$$

follows with the embedding $C^{0,\sigma}(\bar{\Omega}) \hookrightarrow V_\beta^{0,p}(\Omega)$. Inequality (5.69) yields then the assertion (5.76). $\quad\square$

Lemma 5.31. *Let $\bar{u}$ be the solution of the optimal control problem (5.33) with the state equation (4.43) with Neumann boundary conditions (4.46). For the associated state $\bar{y} = S\bar{u}$ and adjoint state $\bar{p} = S\bar{u}$ one has $\bar{y} \in W_\beta^{2,2}(\Omega)$ and $\bar{p} \in W_\beta^{2,2}(\Omega)$ for $\beta > 1 - \pi/\omega$. The a priori estimates*

$$\|\bar{y}\|_{W_\beta^{2,2}(\Omega)} \leq c\|\bar{u}\|_{L^\infty(\Omega)}, \tag{5.77}$$

$$\|\bar{p}\|_{W_\beta^{2,2}(\Omega)} \leq c\left(\|\bar{u}\|_{L^\infty(\Omega)} + \|y_d\|_{C^{0,\sigma}(\bar{\Omega})}\right) \tag{5.78}$$

hold.

Proof. From the computation

$$\|u\|_{W_\beta^{0,2}(\Omega)}^2 = \int_\Omega r^{2\beta} |u|^2 \, \mathrm{d}x \leq \|u\|_{L^\infty(\Omega)}^2 \int_\Omega r^{2\beta} \, \mathrm{d}x$$

and the fact that the last integral is finite due to $2\beta > 2 - 2\pi/\omega > 0$ one can conclude the embedding $L^\infty(\Omega) \hookrightarrow W_\beta^{0,2}(\Omega)$. With this embedding at hand we can apply Lemma 4.16 and get from $\bar{u} \in L^\infty(\Omega)$ the regularity $\bar{y} \in W_\beta^{2,2}(\Omega)$ and the a priori estimate (5.77). From Lemma 5.28 we know that $\bar{y} \in C^{0,\sigma}(\bar{\Omega})$ for $\sigma \in (0, \lambda)$ and therefore $\bar{y} - y_d \in C^{0,\sigma}(\bar{\Omega})$. Since $C^{0,\sigma}(\bar{\Omega}) \hookrightarrow W_\beta^{0,2}(\Omega)$ it follows again from Lemma 4.16 the inequality

$$\|\bar{p}\|_{W_\beta^{2,2}(\Omega)} \leq c\|\bar{y} - y_d\|_{W_\beta^{0,2}(\Omega)} \leq c\left(\|\bar{y} - y_d\|_{C^{0,\sigma}(\bar{\Omega})}\right).$$

Applying the triangle inequality and using estimate (5.69) yields the assertion (5.78). □

5.2.2.2 Approximation error estimate in $L^2(\Omega)$

In this subsection we assume that the prismatic domain Ω is discretized by an anisotropic graded tetrahedral mesh as described in (2.13). The state and adjoint state equation are discretized by a finite element scheme as introduced in Subsection 4.2.2. This means, the approximate state $y_h = S_h u$ is the unique solution of

$$\begin{cases} a_D(y_h, v_h) = (u, v_h)_{L^2(\Omega)} & \forall v_h \in V_{0h} \text{ for state equation (4.45)}, \\ a_N(y_h, v_h) = (u, v_h)_{L^2(\Omega)} & \forall v_h \in V_h \text{ for state equation (4.47)} \end{cases}$$

with

$$V_{0h} = \left\{v \in C(\bar{\Omega}) : v|_T \in \mathcal{P}_1 \ \forall T \in T_h \text{ and } v_h = 0 \text{ on } \partial\Omega\right\},$$
$$V_h = \left\{v \in C(\bar{\Omega}) : v|_T \in \mathcal{P}_1 \ \forall T \in T_h\right\}.$$

Similarly, the approximated adjoint state $p_h = S_h^*(y - y_d)$ is the unique solution of

$$\begin{cases} a_D(v_h, p_h) = (y - y_d, v_h)_{L^2(\Omega)} & \forall v_h \in V_{0h} \text{ for state equation (4.45)}, \\ a_N(v_h, p_h) = (y - y_d, v_h)_{L^2(\Omega)} & \forall v_h \in V_h \text{ for state equation (4.47)}. \end{cases}$$

As in the sections before, we denote by $P_h u = S_h^*(S_h u - y_d)$ the affine operator that maps a given control u to the corresponding approximate adjoint state p_h.

Our aim is to prove that Theorem 5.7 and Theorem 5.14 hold in this situation. To this end we check the Assumptions VAR1 and VAR2 and PP1–PP4 from subsection 5.1.2. Notice, that we have in the notation of that subsection $U = Z = L^2(\Omega)$, $Y = H_0^1(\Omega)$ in case of Dirichlet conditions (4.44) and $Y = H^1(\Omega)$ in case of a Neumann boundary (4.46). Furthermore, we have $X = C^{0,\sigma}(\bar{\Omega})$.

Since the boundedness of the operators S_h and S_h^* play a role in both discretization concepts, the variational discrete approach as well as the postprocessing result, we first repeat the following lemma. Due to the fact that we do not operate on quasi-uniform meshes the boundedness of the operator S_h is not obvious. The following lemma is proved in [127, Subsection 3.6] by using Green function techniques.

Lemma 5.32. *Let T_h be an anisotropic, graded mesh of a prismatic domain with parameter $\mu < \lambda$ according to (2.13). The norms of the discrete solution operators S_h and S_h^* are bounded,*

$$\|S_h\|_{L^2(\Omega)\to L^\infty(\Omega)} \le c, \qquad\qquad \|S_h^*\|_{L^2(\Omega)\to L^\infty(\Omega)} \le c,$$
$$\|S_h\|_{L^2(\Omega)\to L^2(\Omega)} \le c, \qquad\qquad \|S_h^*\|_{L^2(\Omega)\to L^2(\Omega)} \le c,$$
$$\|S_h\|_{L^2(\Omega)\to Y(\Omega)} \le c, \qquad\qquad \|S_h^*\|_{L^2(\Omega)\to Y(\Omega)} \le c,$$

where c is independent of h.

Variational discrete approach   Let us first consider the variational discrete approach that has been introduced in Subsection 5.1.2.1.

Theorem 5.33. *Let $\bar{u}$ be the solution of the optimal control problem (5.33) with the state equation (4.43) with Dirichlet (4.44) or Neumann boundary conditions (4.46) and $\bar{u}_h^s$ the solution of the corresponding variational discrete problem (5.8) on a mesh of type (2.13) with grading parameter $\mu < \lambda$. With the associated states $\bar{y} = S\bar{u}$, $\bar{y}_h^s = S_h \bar{u}_h^s$ and adjoint states $\bar{p} = P\bar{u}$, $\bar{p}_h^s = P_h \bar{u}_h^s$ the estimates*

$$\|\bar{u} - \bar{u}_h^s\|_{L^2(\Omega)} \le ch^2 \left(\|\bar{u}\|_{L^2(\Omega)} + \|y_d\|_{L^2(\Omega)}\right),$$
$$\|\bar{y} - \bar{y}_h^s\|_{L^2(\Omega)} \le ch^2 \left(\|\bar{u}\|_{L^2(\Omega)} + \|y_d\|_{L^2(\Omega)}\right),$$
$$\|\bar{p} - \bar{p}_h^s\|_{L^2(\Omega)} \le ch^2 \left(\|\bar{u}\|_{L^2(\Omega)} + \|y_d\|_{L^2(\Omega)}\right)$$

hold with a constant c independent of h.

Proof. Lemma 5.32 proves Assumption VAR1, Theorem 4.23 and Theorem 4.25 prove Assumption VAR2. Notice, that here $S_h = S_h^*$. The application of Theorem 5.7 yields the assertion of this theorem. $\square$

Postprocessing approach Now we are interested in the postprocessing approach introduced in Subsection 5.1.2.2. In detail, the approximate optimal control $\bar{u}_h$ is the solution of (5.12) and the improved approximate optimal control $\tilde{u}_h$ the projection of scaled discrete adjoint state in the set of admissible controls. As Assumption PP1 is already proved in Lemma 5.32 and the finite element error estimates of Assumption PP2 are shown in Theorems 4.23 and 4.25 we continue with the proof of Assumptions PP3 and PP4. We first concentrate on Assumption PP3. For that proof we give a number of auxiliary results before the assumption can be shown in Lemma 5.39. Some of them were already stated in [127] and we refer for proofs to this thesis. We give proofs here only in those cases when changes are necessary due to the weaker mesh condition $\mu < \lambda$ in comparison with $\mu < \min\{\lambda, \frac{\lambda}{3} + \frac{5}{9}\}$ in [127], or when the proof in [127] is restricted to an analogy argument to a further result.

Before we recall the approximation properties of the operators R_h and Q_h defined in (5.15) and (5.16), we introduce the sets

$$K_s = \bigcup_{\{T \in T_h : r_T = 0\}} T \quad \text{and} \quad K_r = \Omega \backslash \bar{K}_s. \tag{5.79}$$

Notice, that according to (2.13) the number N of elements in K_s is $O(h^{-1})$ and therefore $|K_s| \le cNh^{2/\mu+1} = ch^{2/\mu}$.

Lemma 5.34. *[127, Lemma 3.24] Let T_h be a conforming anisotropic triangulation satisfying equation (2.13). Then there holds*

$$\left| \int_T (f - R_h f) \, dx \right| \le \begin{cases} c|T|^{1/2} \sum_{|\alpha|=2} h_T^\alpha \|D^\alpha f\|_{L^2(T)} & \text{for } f \in H^2(T) \\ c|T| \sum_{|\alpha|=1} h_T^\alpha \|D^\alpha f\|_{L^\infty(T)} & \text{for } f \in W^{1,\infty}(T) \\ c|T| \|f\|_{L^\infty(T)} & \text{for } f \in L^\infty(T). \end{cases}$$

Lemma 5.35. *[127, Lemma 4.13] The inequality*

$$\|Q_h f - R_h f\|_{L^2(T)} \le |T|^{1/2-1/p} \sum_{|\alpha|=1} h_T^\alpha \|D^\alpha f\|_{L^p(T)}$$

holds for all $f \in W^{1,p}(T)$ with $p > 3$.

Proof. By the definition of Q_h and R_h one has

$$\int_T (Q_h f - R_h f)^2 \, dx = \int_T \left[\frac{1}{|T|} \int_T f - R_h f d\xi \right]^2 dx = |T|^{-1} \left[\int_T f - R_h f d\xi \right]^2$$

which leads to

$$\|Q_h f - R_h f\|_{L^2(T)} \le |T|^{-1/2} \left| \int_T f - R_h f \, dx \right|. \tag{5.80}$$

For any $\hat{w} \in \mathcal{P}_0(\hat{T})$ we can conclude

$$\int_T (f - R_h f) \, dx = |T| \int_{\hat{T}} (\hat{f} - \hat{R}\hat{f}) \, dx = |T| \int_{\hat{T}} (\hat{f} - \hat{w}) - \hat{R}(\hat{f} - \hat{w}) \, dx$$

$$\leq c|T| \|\hat{f} - \hat{w}\|_{L^\infty(\hat{T})} \leq c|T| \|\hat{f} - \hat{w}\|_{W^{1,p}(\hat{T})}$$

where we have used the embedding $W^{1,p}(\hat{T}) \hookrightarrow L^\infty(\hat{T})$ for $p > 3$. Now we can apply the Deny-Lions lemma and get

$$\int_T (f - R_h f) \, dx \leq c|T| \|\hat{f}\|_{W^{1,p}(\hat{T})} \leq c|T|^{1-1/p} \sum_{|\alpha|=1} h_T^\alpha \|D^\alpha f\|_{L^p(T)}$$

which, together with estimate (5.80), yields the assertion. $\qquad\square$

Corollary 5.36. *[127, Corollary 4.16] Let the mesh be graded according to (2.13). Then the estimate*

$$\|Q_h w - R_h w\|_{L^2(K_s)} \leq ch^2 \left(\|\partial_1 w\|_{L^p(K_s)} + \|\partial_2 w\|_{L^p(K_s)} + \|r^{-\mu}\partial_3 w\|_{L^p(K_s)} \right)$$

holds for all $w \in W^{1,p}(K_s)$ with $r^{-\mu}\partial_3 w \in L^p(K_s)$ and $p > 3$, $p \geq \frac{1}{1-\mu}$.

Corollary 5.37. *Let the mesh be graded according to (2.13). Then the estimate*

$$\|Q_h w - R_h w\|_{L^2(K_r)} \leq ch^2 \left(|w|_{V^{2,2}_{\frac{3}{2}-2\mu}(K_r)} + |\partial_3 w|_{V^{2,1}_{1-\mu}(K_r)} + \|\partial_{33} w\|_{L^2(K_r)} \right)$$

holds for all $w \in H^2(K_r)$.

Proof. The proof is taken from [127, pages 48f. and 23]. From the definition of Q_h one has

$$\|Q_h w - R_h w\|^2_{L^2(K_r)} = \sum_{T \subset K_r} \|Q_h w - R_h w\|^2_{L^2(T)}$$

$$= \sum_{T \subset K_r} |T|^{-1} \left| \int_T (w - R_h w) \, dx \right|^2$$

We apply Lemma 5.34 and get

$$\|Q_h w - R_h w\|^2_{L^2(K_r)} \leq \sum_{T \subset K_r} |T|^{-1} \left[c \sum_{|\alpha|=2} h_T^\alpha |T|^{1/2} \|D^\alpha w\|_{L^2(T)} \right]^2$$

$$\leq c \sum_{T \subset K_r} \left[\sum_{|\alpha|=2} h_T^\alpha \|D^\alpha w\|_{L^2(T)} \right]^2.$$

Since $r_T > 0$ for an element $T \subset K_r$ it follows with (2.13)

$$\|Q_h w - R_h w\|^2_{L^2(K_r)} \le ch^4 \sum_{T \subset K_r} \left[r_T^{2-2\mu} \sum_{i=1}^{2} \sum_{j=1}^{2} \|\partial_{ij} w\|_{L^2(T)} \right.$$

$$\left. + r_T^{1-\mu} \sum_{i=1}^{2} \|\partial_{3i} w\|_{L^2(T)} + \|\partial_{33} w\|_{L^2(T)} \right]^2$$

$$\le ch^4 \left(|w|_{V_{2-2\mu}^{2,2}(K_r)} + |\partial_3 w|_{V_{1-\mu}^{1,2}(K_r)} + \|\partial_{33} w\|_{L^2(K_r)} \right)^2.$$

Extracting the root yields the assertion. $\qquad\square$

The following lemma includes a stronger result in comparison with [127, Corollary 4.16].

Corollary 5.38. *Let the mesh be graded according to (2.13). Then the estimate*

$$\|Q_h w - R_h w\|_{L^2(K_s)} \le ch^2 \|\nabla w\|_{L^\infty(K_s)}$$

holds for all $w \in W^{1,\infty}(K_s)$.

Proof. One can conclude from the Definition of Q_h and Lemma 5.34

$$\|Q_h w - R_h w\|^2_{L^2(K_s)} = \sum_{T \subset K_s} \int_T \left[|T|^{-1} \int_T w - R_h w \, d\xi \right]^2 dx$$

$$= \sum_{T \subset K_s} |T|^{-1} \left[\int_T w - R_h w \, d\xi \right]^2$$

$$\le c \sum_{T \subset K_s} |T| \left(\sum_{|\alpha|=1} h_T^\alpha \|D^\alpha w\|_{L^\infty(T)} \right)^2.$$

If one takes into account that $\#K_s \le ch^{-1}$ it follows

$$\|Q_h w - R_h w\|^2_{L^2(K_s)} \le ch^{3+2/\mu} \#K_s \|\nabla w\|^2_{L^\infty(K_s)} \le ch^{2+2/\mu} \|\nabla w\|^2_{L^\infty(K_s)}.$$

Since $\mu \le 1$ this yields the assertion. $\qquad\square$

Now we are able to prove Assumption PP3 in the following lemma.

Lemma 5.39. *Let the mesh be graded according to (2.13) with $\mu < \lambda$. Then the inequality*

$$\|Q_h \bar{p} - R_h \bar{p}\|_{L^2(\Omega)} \le ch^2 \left(\|\bar{u}\|_{C^{0,\sigma}(\bar{\Omega})} + \|y_d\|_{C^{0,\sigma}(\bar{\Omega})} \right)$$

holds.

Proof. We write

$$\|R_h\bar{p} - Q_h\bar{p}\|^2_{L^2(\Omega)} = \|R_h\bar{p} - Q_h\bar{p}\|^2_{L^2(K_s)} + \|R_h\bar{p} - Q_h\bar{p}\|^2_{L^2(K_r)}. \tag{5.81}$$

with K_s and K_r as defined in (5.79). In the following we choose p such that $p > 3$ and $p \geq \frac{1}{1-\mu}$. According to (4.49) we split $\bar{p}$ in a singular part $\bar{p}_s$ and a regular part $\bar{p}_r \in W^{2,p}(K_s)$ such that $\bar{p} = \bar{p}_s + \bar{p}_r$. For the singular part we get from (5.81) and the Corollaries 5.36 and 5.37 the estimate

$$\|R_h\bar{p}_s - Q_h\bar{p}_s\|_{L^2(\Omega)} \leq ch^2 \left(|\bar{p}_s|_{V^{2,2}_{2-2\mu}(K_r)} + |\partial_3\bar{p}_s|_{V^{1,2}_{1-\mu}(K_r)} + |\partial_{33}\bar{p}_s|_{V^{0,2}_0(K_r)} \right.$$
$$\left. + \|\partial_1\bar{p}_s\|_{L^p(K_s)} + \|\partial_2\bar{p}_s\|_{L^p(K_s)} + \|\partial_3\bar{p}_s\|_{V^{0,p}_{-\mu}(K_s)} \right)$$
$$\leq ch^2\|\bar{y} - y_d\|_{L^p(\Omega)} \tag{5.82}$$

where we have used the estimates (4.50)–(4.52) in the last step. Since $W^{2,p}(\Omega) \hookrightarrow H^2(\Omega)$ and $W^{2,p}(\Omega) \hookrightarrow W^{1,\infty}(\Omega)$ (comp. Theorem 2.13) it follows from (5.81) and the Corollaries 5.37 and 5.38

$$\|R_h\bar{p}_r - Q_h\bar{p}_r\|_{L^2(\Omega)}$$
$$\leq ch^2 \left(|\bar{p}_r|_{V^{2,2}_{2-2\mu}(K_r)} + |\partial_3\bar{p}_r|_{V^{1,2}_{1-\mu}(K_r)} + |\partial_{33}\bar{p}_r|_{V^{0,2}_0(K_r)} + \|\nabla\bar{p}_r\|_{L^\infty(K_s)} \right)$$
$$\leq ch^2\|\bar{y} - y_d\|_{L^p(\Omega)} \tag{5.83}$$

where we have used in the last step the a priori estimates of Lemma 4.15 and 4.16, respectively, and the estimate (4.53). Since

$$\|R_h\bar{p} - Q_h\bar{p}\|_{L^2(\Omega)} \leq \|R_h\bar{p}_s - Q_h\bar{p}_s\|_{L^2(\Omega)} + \|R_h\bar{p}_r - Q_h\bar{p}_r\|_{L^2(\Omega)}$$

one can conclude from (5.82) and (5.83)

$$\|R_h\bar{p} - Q_h\bar{p}\|_{L^2(\Omega)} \leq ch^2\|\bar{y} - y_d\|_{L^p(\Omega)}.$$

Finally it follows from the triangle inequality, the embedding $C^{0,\sigma}(\bar{\Omega}) \hookrightarrow L^p(\Omega)$ and Lemma 5.28

$$\|R_h\bar{p} - Q_h\bar{p}\|_{L^2(\Omega)} \leq ch^2 \left(\|\bar{u}\|_{C^{0,\sigma}(\bar{\Omega})} + \|y_d\|_{C^{0,\sigma}(\bar{\Omega})} \right)$$

what is the assertion of this lemma. $\square$

It remains the proof of Assumption PP4. The proof of this assumption uses the boundedness of $r^\beta\nabla\bar{p}$ for $\beta > 1 - \lambda$ stated in Corollary 5.29. In [127] this boundedness was only proven for $\beta > \frac{4}{3} - \lambda$. Our improvement allows us to weaken the grading condition from $\mu < \min\{\lambda, 5/9 + \lambda/3\}$ as it is given in [127] to $\mu < \lambda$. Notice, that the condition $\mu < \lambda$ was also necessary to get optimal convergence of the finite element approximation of the state equation (comp. Theorems 4.23 and 4.25 and also [6]).

From the projection formula (5.6) one can see, that there may be elements where the optimal control $\bar{u}$ admits kinks. For such an element T one cannot assume that the restriction $\bar{u}|_T$ is contained in $V_\beta^{2,2}(T)$. Consequently, a special treatment is necessary during the error analysis. Therefore we split the domain Ω in two parts,

$$K_1 := \bigcup_{T \in \mathcal{T}_h : \bar{u} \notin V_\beta^{2,2}(T)} T, \quad K_2 := \bigcup_{T \in \mathcal{T}_h : \bar{u} \in V_\beta^{2,2}(T)} T. \tag{5.84}$$

Clearly, the number of elements in K_1 grows for decreasing h. Nevertheless, it is quite reasonable to assume that the boundary of the active set has finite two-dimensional measure, i.e.

$$|K_1| \leq ch. \tag{5.85}$$

Notice, that this is a weaker condition than $\#K_1 \leq ch^{-2}$ as it is required in [127]. For a detailed discussion on this, we refer to [127, Lemma 4.7].

Lemma 5.40. *Let $\mathcal{T}_h$ be an anisotropic, graded mesh satisfying (2.13) with $\mu < \lambda$. Let $\bar{u}$ be the solution of the optimal control problem (5.33) with the state equation (4.45) or (4.47). Then the estimate*

$$(Q_h\bar{u} - R_h\bar{u}, v_h)_{L^2(\Omega)} \leq ch^2\|v_h\|_{L^\infty(\Omega)} \left(\|\bar{u}\|_{L^\infty(\Omega)} + \|\bar{y}_d\|_{C^{0,\sigma}(\bar{\Omega})}\right)$$

holds for all $v_h \in V_h$.

Proof. We follow the lines of the proof of Lemma 4.10 in [127]. Since the mesh grading condition is weakened from $\mu < \min\{\lambda, \frac{5}{9} + \frac{\lambda}{3}\}$ to $\mu < \lambda$ a detailed proof is given. Furthermore, modifications are necessary to cover also the Neumann case. We split the domain Ω into three parts, where $\bar{u}$ has different regularity, $K_{1,r} = K_1 \backslash \bar{K}_s$, $K_{2,r} = K_2 \backslash \bar{K}_s$ and K_s. One has

$$\int_\Omega v_h(Q_h\bar{u} - R_h\bar{u}) \, \mathrm{d}x \leq \sum_{T \in \mathcal{T}_h} \|v_h\|_{L^\infty(T)} \int_T (\bar{u} - R_h\bar{u}) \, \mathrm{d}x.$$

If we apply Lemma 5.34 on each sub-domain to the integral, we get

$$\begin{aligned}
\int_\Omega v_h(Q_h\bar{u} - R_h\bar{u}) \, \mathrm{d}x \leq &\sum_{T \subset K_{2,r}} \|v_h\|_{L^\infty(T)} |T|^{1/2} \sum_{|\alpha|=2} h_T^\alpha \|D^\alpha \bar{u}\|_{L^2(T)} \\
&+ \sum_{T \subset K_{1,r}} \|v_h\|_{L^\infty(T)} |T| \sum_{|\alpha|=1} h_T^\alpha \|D^\alpha \bar{u}\|_{L^\infty(T)} \\
&+ \sum_{T \subset K_s} \|v_h\|_{L^\infty(T)} |T| \|\bar{u}\|_{L^\infty(T)}.
\end{aligned} \tag{5.86}$$

We estimate the three terms on the right-hand side separately using (2.13). For the first term we have

$$\sum_{T \subset K_{2,r}} \|v_h\|_{L^\infty(T)} |T|^{1/2} \sum_{|\alpha|=2} h_T^\alpha \|D^\alpha \bar{u}\|_{L^2(T)}$$

$$\leq c \|v_h\|_{L^\infty(K_{2,r})} |K_{2,r}|^{1/2} \left(h^2 \sum_{i=1}^{2} \sum_{j=1}^{2} \|r^{2-2\mu} \partial_{ij} \bar{u}\|_{L^2(K_{2,r})} \right.$$

$$\left. + h^2 \sum_{i=1}^{2} \|r^{1-\mu} \partial_{3i} \bar{u}\|_{L^2(K_{2,r})} + h^2 \|\partial_{33} \bar{u}\|_{L^2(K_{2,r})} \right). \tag{5.87}$$

The second term can be estimated by using (2.13),

$$\sum_{T \subset K_{1,r}} \|v_h\|_{L^\infty(T)} |T| \sum_{|\alpha|=1} h_T^\alpha \|D^\alpha \bar{u}\|_{L^\infty(T)}$$

$$\leq ch \|v_h\|_{L^\infty(\Omega)} \sum_{T \subset K_{1,r}} |T| \left(\sum_{i=1}^{2} \|r^{1-\mu} \partial_i \bar{u}\|_{L^\infty(T)} + \|\partial_3 \bar{u}\|_{L^\infty(T)} \right)$$

$$\leq ch \|v_h\|_{L^\infty(\Omega)} |K_{1,r}| \left(\sum_{i=1}^{2} \|r^{1-\mu} \partial_i \bar{u}\|_{L^\infty(K_{1,r})} + \|\partial_3 \bar{u}\|_{L^\infty(K_{1,r})} \right)$$

$$\leq ch^2 \|v_h\|_{L^\infty(\Omega)} \left(\sum_{i=1}^{2} \|r^{1-\mu} \partial_i \bar{u}\|_{L^\infty(K_{1,r})} + \|\partial_3 \bar{u}\|_{L^\infty(K_{1,r})} \right). \tag{5.88}$$

The last step is valid since $|K_1| \leq ch$ (comp. (5.85)). The third term yields

$$\sum_{T \subset K_s} \|v_h\|_{L^\infty(T)} |T| \|\bar{u}\|_{L^\infty(T)} \leq |K_s| \|v_h\|_{L^\infty(\Omega)} \|\bar{u}\|_{L^\infty(K_s)}$$

$$\leq ch^2 \|v_h\|_{L^\infty(\Omega)} \|\bar{u}\|_{L^\infty(K_s)} \tag{5.89}$$

since $|K_s| \leq ch^{2/\mu} \leq ch^2$. We can further utilize the projection formula (5.6) and substitute $\bar{u}$ by $-\frac{1}{\nu}\bar{p}$ in the above norms, because $\bar{u}$ is either constant or equal to $-\frac{1}{\nu}\bar{p}$. Then the inequalities (5.87), (5.88) and (5.89) yield together with (5.86) the estimate

$$\int_\Omega v_h (Q_h \bar{u} - R_h \bar{u}) \, dx \leq \frac{c}{\nu} h^2 \|v_h\|_{L^\infty(\Omega)}.$$

$$\left(\sum_{i=1}^{2} \sum_{j=1}^{2} \|r^{2-2\mu} \partial_{ij} \bar{p}\|_{L^2(K_{2,r})} + \sum_{i=1}^{2} \|r^{1-\mu} \partial_{3i} \bar{p}\|_{L^2(K_{2,r})} + \|\partial_{33} \bar{p}\|_{L^2(K_{2,r})} \right. \tag{5.90}$$

$$\left. + \sum_{i=1}^{2} \|r^{1-\mu} \partial_i \bar{p}\|_{L^\infty(K_{1,r})} + \|\partial_3 \bar{p}\|_{L^\infty(K_{1,r})} + \nu \|\bar{u}\|_{L^\infty(K_s)} \right). \tag{5.91}$$

In order to estimate the L^2-norms in (5.90), we split $\bar{p}$ according to (4.49) in a regular and a singular part, $\bar{p} = \bar{p}_r + \bar{p}_s$. Then we apply Lemma 4.17 for $p = 2$ with $\beta = 2 - 2\mu$.

This is possible since $\mu < \lambda < 1$ and therefore $2 - 2\mu > 2 - \lambda - 1 = 1 - \lambda$. Notice that one has for the regular part $\|r^\alpha \bar{p}_r\|_{H^2(\Omega)} \leq c\|\bar{p}_r\|_{H^2(\Omega)}$ as long as $\alpha > 0$. For the estimate of the L^∞-norms in (5.91), we apply Corollary 5.29. We end up with

$$\int_\Omega v_h (Q_h \bar{u} - R_h \bar{u}) \, dx \leq$$

$$ch^2 \|v_h\|_{L^\infty(\Omega)} \left(\|\bar{y} - y_d\|_{L^2(\Omega)} + \|\bar{y} - y_d\|_{C^{0,\sigma}(\bar{\Omega})} + \|\bar{u}\|_{L^\infty(\Omega)} \right)$$

for a $\sigma \in (0,1)$. If we use the triangle inequality, the embedding $C^{0,\sigma}(\bar{\Omega}) \hookrightarrow L^\infty(\Omega) \hookrightarrow L^2(\Omega)$ and estimate (5.69) the assertion is shown. $\square$

Now we are ready to summarize the results in the following theorem.

Theorem 5.41. *Let $\bar{u}$ be the solution of the optimal control problem (5.33) with the state equation (4.45) or (4.47) and $\bar{u}_h$ the corresponding discrete solution of (5.12) on a mesh of type (2.13) with $\mu < \lambda$. Furthermore, let $\bar{y} = S\bar{u}$, $\bar{p} = P\bar{u}$, $\bar{y}_h = S_h \bar{u}_h$, $\bar{p}_h = P_h \bar{u}_h$ be the associated states and adjoint states and $\tilde{u}_h$ be the postprocessed control constructed by (5.14). Then the estimates*

$$\|\bar{y} - \bar{y}_h\|_{L^2(\Omega)} \leq ch^2 \left(\|\bar{u}\|_{C^{0,\sigma}(\bar{\Omega})} + \|y_d\|_{C^{0,\sigma}(\bar{\Omega})} \right)$$

$$\|\bar{p} - \bar{p}_h\|_{L^2(\Omega)} \leq ch^2 \left(\|\bar{u}\|_{C^{0,\sigma}(\bar{\Omega})} + \|y_d\|_{C^{0,\sigma}(\bar{\Omega})} \right)$$

$$\|\bar{u} - \tilde{u}_h\|_{L^2(\Omega)} \leq ch^2 \left(\|\bar{u}\|_{C^{0,\sigma}(\bar{\Omega})} + \|y_d\|_{C^{0,\sigma}(\bar{\Omega})} \right)$$

hold true.

Proof. The assertion follows from Theorem 5.14. Assumption PP1 is proved in Lemma 5.32, Assumption PP2 in Theorem 4.23 and 4.25, respectively, Assumption PP3 in Lemma 5.39 and Assumption PP4 in Lemma 5.40. $\square$

Remark 5.42. Winkler proved in his thesis [127] second order convergence for a special type of mixed boundary condition under the stronger mesh grading condition $\mu < \min\{\lambda, 5/9 + \lambda/3\}$. For a discussion on the grading condition we refer to page 115 and also [127, Remark 4.11].

Remark 5.43. Apel and Winkler proved in [19] the result of Theorem 5.14 for domains with corner- and edge singularities and appropriately graded isotropic meshes. In detail, the mesh was chosen such that the condition

$$h_T \sim h^{1/\mu} \text{ for } r_T = 0,$$

$$h_T \sim h r_T^{1-\mu} \text{ for } r_T > 0$$

is satisfied, where h_T denotes the diameter of the element T and r_T its distance to the set of singular points. The grading parameter μ had to fulfill the three conditions

$$\mu < \frac{1}{2} + \frac{1}{2}\lambda_v, \quad \mu < \lambda_e, \quad \mu < \frac{1}{3} + \frac{1}{2}\lambda_e. \tag{5.92}$$

Here λ_v and λ_e denote particular eigenvalues of certain operator pencils that correspond to the corner- and edge singularities, respectively. As in the case of anisotropic refinement, a weaker condition, namely

$$\mu < \min\left\{\frac{1}{2} + \lambda_v, \lambda_e\right\} \tag{5.93}$$

is sufficient to get an optimal convergence rate for the boundary value problem [11]. Let us quickly describe where the additional conditions $\mu < \frac{1}{3} + \frac{1}{2}\lambda_e$ and $\mu < \frac{1}{2} + \frac{1}{2}\lambda_v$ come from in [19]. In that paper the boundedness for $r^\beta \nabla p$ was proved for

$$\beta > \max\left\{\frac{4}{3} - \lambda_e, 1 - \lambda_v\right\} \tag{5.94}$$

by the use of Sobolev embedding theorems. In the proof of Lemma 4.5 in [19] one needed the boundedness of $r^{2-2\mu}\nabla p$. This resulted in the condition $2 - 2\mu > \max\{\frac{4}{3} - \lambda_e, 1 - \lambda_v\}$, i.e. $\mu < \min\{\frac{1}{3} + \frac{1}{2}\lambda_e, \frac{1}{2} + \frac{1}{2}\lambda_v\}$. With an analogous argumentation as in Lemma 4.20 and Corollary 5.29 one can prove that $r^\beta \nabla p$ is already bounded for

$$\beta > \max\left\{1 - \lambda_e, 1 - \lambda_v\right\}$$

as long as the desired state is in $C^{0,\sigma}(\bar{\Omega})$. This means a smaller weight than stated in (5.94) is sufficient to compensate a possible edge singularity. Consequently, the condition in the proof of Lemma 4.5 in [19] reduces to $2 - 2\mu > \max\{1 - \lambda_e, 1 - \lambda_v\}$, what is fulfilled by values of μ that satisfy $\mu < \min\{\frac{1}{2} + \frac{1}{2}\lambda_e, \frac{1}{2} + \frac{1}{2}\lambda_v\}$. Since $\lambda_e < 1$ the condition $\mu < \frac{1}{2} + \frac{1}{2}\lambda_e$ is weaker than $\mu < \lambda_e$. Therefore one gets second order convergence on isotropic graded meshes already for a grading parameter μ satisfying

$$\mu < \min\left\{\frac{1}{2} + \frac{1}{2}\lambda_v, \lambda_e\right\}$$

what is of course a weaker condition than the original condition (5.92). Notice, that this condition is still slightly stronger than condition (5.93).

5.2.2.3 Numerical tests

We illustrate the theoretical findings of this subsection by two numerical examples.

Dirichlet boundary conditions   We consider the optimal control problem (5.33) with the state equation

$$-\Delta y = u + f \text{ in } \Omega, \qquad y = 0 \text{ on } \partial\Omega.$$

The domain Ω is chosen as

$$\Omega - \{(r\cos\varphi, r\sin\varphi, z) \subset \mathbb{R}^3 : 0 < r < 1, 0 < \varphi < \omega_0, 0 < z < 1\}.$$

Table 5.3: $L^2(\Omega)$-error of the computed control $\tilde{u}_h$, state $\bar{y}_h$ and adjoint state $\bar{p}_h$ for Dirichlet boundary conditions on anisotropic graded meshes ($\mu = 0.4$)

ndof	$\|\bar{u} - \tilde{u}_h\|$	eoc	$\|\bar{y} - \bar{y}_h\|$	eoc	$\|\bar{p} - \bar{p}_h\|$	eoc
224	2.31e−01		1.31e−02		2.58e−04	
2349	6.58e−02	1.60	3.72e−03	1.60	8.89e−05	1.36
21299	1.74e−02	1.81	1.08e−03	1.68	3.47e−05	1.28
74849	7.80e−03	1.91	5.19e−04	1.76	1.79e−05	1.59
180999	4.37e−03	1.96	2.99e−04	1.87	1.05e−05	1.81
357749	2.79e−03	1.97	1.94e−04	1.92	6.84e−06	1.88
623099	1.94e−03	1.98	1.35e−04	1.94	4.80e−06	1.91
995049	1.42e−03	1.98	9.98e−05	1.96	3.55e−06	1.94

Table 5.4: $L^2(\Omega)$-error of the computed control $\tilde{u}_h$, state $\bar{y}_h$ and adjoint state $\bar{p}_h$ for Dirichlet boundary conditions on quasi-uniform meshes ($\mu = 1.0$)

ndof	$\|\bar{u} - \tilde{u}_h\|$	eoc	$\|\bar{y} - \bar{y}_h\|$	eoc	$\|\bar{p} - \bar{p}_h\|$	eoc
224	1.37e−01		6.28e−03		1.64e−04	
2349	3.71e−02	1.67	1.96e−03	1.48	6.13e−05	1.26
21299	1.21e−02	1.52	7.97e−04	1.23	2.86e−05	1.04
74849	6.42e−03	1.52	4.76e−04	1.23	1.67e−05	1.29
180999	4.17e−03	1.47	3.31e−04	1.23	1.13e−05	1.34
357749	3.02e−03	1.42	2.52e−04	1.21	8.30e−06	1.34
623099	2.34e−03	1.38	2.02e−04	1.19	6.50e−06	1.32
995049	1.89e−03	1.37	1.68e−05	1.19	5.30e−06	1.32

The functions f and y_d are defined such that

$$\bar{y}(r, \varphi, z) = z(1 - z)(r^\lambda - r^\alpha) \sin \lambda\varphi,$$
$$\bar{p}(r, \varphi, z) = \nu z(1 - z)(r^\lambda - r^\alpha) \sin \lambda\varphi,$$
$$\bar{u}(r, \varphi, z) = \Pi_{[-0.05, 10.0]}\left(-\frac{1}{\nu}\bar{p}\right)$$

is the exact solution of the optimal control problem. We set $\omega_0 = \frac{11}{6}\pi$, $\nu = 10^{-3}$ and $\alpha = \frac{5}{2}$. Furthermore, we have $\lambda = \frac{\pi}{\omega} = \frac{6}{11}$.

In Table 5.3 one can find the values for the errors $\|\bar{u} - \tilde{u}_h\|_{L^2(\Omega)}$, $\|\bar{y} - \bar{y}_h\|_{L^2(\Omega)}$ and $\|\bar{p} - \bar{p}_h\|_{L^2(\Omega)}$ as well as the estimated rates of convergence for different numbers of degrees of freedom on an anisotropic graded mesh with $\mu = 0.4$. One can observe the predicted convergence rate of 2 in all three variables. Table 5.4 shows the values for quasi-uniform meshes ($\mu = 1$). The convergence rates are significantly less than 2, but larger than the rate of $2\lambda = 12/11$ as expected from the theory. However this is an asymptotic result for a region near the edge. Nevertheless, a comparison of both tables reveals a significant improvement of the convergence rates on appropriately graded meshes.

Table 5.5: $L^2(\Omega)$-errors of the computed control $\tilde{u}_h$, state $\bar{y}_h$ and adjoint state $\bar{p}_h$ for Neumann boundary conditions on anisotropic graded meshes ($\mu = 0.4$)

ndof	$\|u - \tilde{u}_h\|$	eoc	$\|y - \bar{y}_h\|$	eoc	$\|p - \bar{p}_h\|$	eoc
576	3.86e−02		4.12e−03		3.62e−04	
3751	1.19e−02	1.88	1.15e−03	2.04	9.33e−05	2.17
26901	3.34e−03	1.94	3.02−04	2.04	2.35e−05	2.10
87451	1.54e−03	1.98	1.36e−04	2.03	1.05e−05	2.05
203401	8.78e−04	1.99	7.70e−05	2.02	5.92e−06	2.04
392751	5.67e−04	1.99	4.94e−05	2.02	3.79e−06	2.03
673501	3.97e−04	1.98	3.44e−05	2.02	2.63e−06	2.03
1063651	2.94e−04	1.98	2.53e−05	2.02	1.94e−06	2.02

Neumann boundary conditions   As second example we consider the optimal control problem (5.33) with the state equation

$$-\Delta y + y = u + f \quad \text{in } \Omega, \qquad \frac{\partial y}{\partial n} = 0 \quad \text{on } \partial\Omega.$$

We choose Ω, ω_0, ν and α like above in the case of Dirichlet boundary conditions. The functions f and y_d are defined such that

$$\bar{y}(r, \varphi, z) = \left(\frac{2}{3}z^3 - z^2\right)(r^\lambda - r^\alpha)\cos\lambda\varphi,$$

$$\bar{p}(r, \varphi, z) = -\nu\left(\frac{2}{3}z^3 - z^2\right)(r^\lambda - r^\alpha)\cos\lambda\varphi,$$

$$\bar{u}(r, \varphi, z) = \Pi_{[-0.05, 10.0]}\left(-\frac{1}{\nu}\bar{p}\right)$$

is the exact solution of the optimal control problem.

Tables 5.5 and 5.6 show the values for the errors $\|\bar{u} - \tilde{u}_h\|_{L^2(\Omega)}$, $\|\bar{y} - \bar{y}_h\|_{L^2(\Omega)}$ and $\|\bar{p} - \bar{p}_h\|_{L^2(\Omega)}$ as well as the estimated rates of convergence for different numbers of degrees of freedom for graded and quasi-uniform meshes, respectively. On the graded meshes ($\mu = 0.4$) one observes a convergence rate of 2 in all three variables as predicted by the theory. On quasi-uniform meshes ($\mu = 1$) the convergence rate in the control is about $2\lambda = 12/11 \approx 1.09$. The rates for state and adjoint state are slightly better but still significantly smaller than two. This means that mesh grading yields also in this case a markable improvement in the error reduction.

5.2.3 Nonsmooth coefficients

In this subsection we consider the optimal control problem (5.33) with state equation (4.79).

Table 5.6: $L^2(\Omega)$-errors of the computed control $\tilde{u}_h$, state $\bar{y}_h$ and adjoint state $\bar{p}_h$ for Neumann boundary conditions on quasi-uniform meshes ($\mu = 1.0$)

ndof	$\|u - \tilde{u}_h\|$	eoc	$\|y - \bar{y}_h\|$	eoc	$\|p - \bar{p}_h\|$	eoc
576	4.61e−02		5.28e−03		4.78e−04	
3751	1.79e−02	1.52	1.82e−03	1.71	1.50e−04	1.86
26901	8.15e−03	1.20	7.50−04	1.35	5.26e−05	1.59
87451	5.27e−03	1.11	4.63e−04	1.23	2.99e−05	1.44
203401	3.88e−03	1.09	3.33e−04	1.18	2.04e−05	1.36
392751	3.06e−03	1.08	2.58e−04	1.16	1.53e−05	1.31
673501	2.52e−03	1.08	2.10e−04	1.14	1.22e−05	1.27
1063651	2.13e−03	1.08	1.77e−05	1.13	1.01e−05	1.25

5.2.3.1 Regularity

As one could already observe for the boundary value problem the regularity properties in the subdomains Ω_i are similar to those of scalar elliptic equations in corner domains. Only the computation of the singular exponent differs. We summarize the regularity in the whole domain Ω in the following lemma.

Lemma 5.44. *For the solution $\bar{u}$ of the optimal control problem (5.33) with state equation (4.79) and the associated state $\bar{y} = S\bar{u}$ and adjoint state $\bar{p} = S\bar{u}$ one has $\bar{u} \in C^{0,\sigma}(\bar{\Omega})$, $\bar{y} \in C^{0,\sigma}(\bar{\Omega}) \cap V_\beta^{2,p}(\Omega)$ and $\bar{p} \in C^{0,\sigma}(\bar{\Omega}) \cap V_\beta^{2,p}(\Omega)$ for some $\sigma \in (0,1]$, for all $p \in (1,\infty)$ and $\beta > 2 - \lambda - 2/p$ with λ according to (4.82). The a priori estimates*

$$\|\bar{y}\|_{C^{0,\sigma}(\bar{\Omega})} + \|\bar{y}\|_{V_\beta^{2,p}(\Omega)} \le c\|\bar{u}\|_{L^\infty(\Omega)} \le c\|\bar{u}\|_{C^{0,\sigma}(\bar{\Omega})}$$

$$\|\bar{p}\|_{C^{0,\sigma}(\bar{\Omega})} + \|\bar{p}\|_{V_\beta^{2,p}(\Omega)} \le c\left(\|\bar{u}\|_{C^{0,\sigma}(\bar{\Omega})} + \|\bar{y}_d\|_{C^{0,\sigma}(\bar{\Omega})}\right)$$

hold for all $p \in (1,\infty)$ and $\beta > 2 - \lambda - 2/p$.

Proof. To prove the assertion one can apply the same argumentation in every subdomain Ω_i, $i = 1,\ldots,n$, as in Lemma 5.16 using Lemma 4.26 instead of Lemma 4.1. □

Corollary 5.45. *For $p < 2/(1 - \lambda)$ the estimate*

$$\|\bar{p}\|_{W^{1,p}(\Omega)} \le c\left(\|\bar{u}\|_{C^{0,\sigma}(\bar{\Omega})} + \|\bar{y}_d\|_{C^{0,\sigma}(\bar{\Omega})}\right)$$

holds. Furthermore, the inequality

$$\|\bar{p}\|_{V_\beta^{1,\infty}(\Omega)} \le c\left(\|\bar{u}\|_{C^{0,\sigma}(\bar{\Omega})} + \|\bar{y}_d\|_{C^{0,\sigma}(\bar{\Omega})}\right)$$

is true for $\beta > 1 - \lambda$.

Proof. For $p < 2/(1 - \lambda)$ one can choose $\beta = 1$ in Lemma 5.44 such that the embedding $V_\beta^{2,p}(\Omega) \hookrightarrow V^{2-\beta,p}(\Omega) \hookrightarrow W^{1,p}(\Omega)$ yields the first assertion. For the second assertion we refer to the proof of Corollary 1 in [15]. The same arguments can be applied in every subdomain Ω_i, $i = 1,\ldots,n$, and the desired inequality follows directly. □

5.2.3.2 Approximation error estimates in $L^2(\Omega)$

Let the domain Ω as introduced in Section 4.3 be discretized by an isotropic graded mesh as described in (2.12). We assume again that the triangulation is aligned with the partition of Ω (comp. Subsection 4.3.2). The approximate state $y_h = S_h u$ is the unique solution of

$$a_I(y_h, v_h) = (u, v_h)_{L^2(\Omega)} \quad \forall v_h \in V_{0h}$$

with bilinear form a_I from (4.81) and V_{0h} defined in (4.83). The approximate adjoint state $p_h = S_h(y - y_d)$ is the unique solution of

$$a_I(p_h, v_h) = (y - y_d, v_h)_{L^2(\Omega)} \quad \forall v_h \in V_{0h}.$$

Again P_h is the affine operator that maps a given control u to the corresponding adjoint state p_h, $P_h u = S_h^*(S_h u - y_d)$.

In the following we show that Assumptions VAR1 and VAR2 as well as PP1–PP4 hold and therefore the error estimates of Theorems 5.7 and 5.14 are valid. Although the proofs are similar to those in the sections above we use this example to illustrate how one can profit from the general formulation in Section 5.1. The assumptions given there allow to extend the results very easily to the case where the interface problem of the Laplacian serves as state equation. The basis has been formed in Section 4.3 with the regularity results and finite element error estimates for the state equation.

Since the regularity parameter λ can be smaller than $1/2$ it is important to guarantee the boundedness of S_h and S_h^* also in this case. The proof of Lemma 5.17 works also for discontinuous coefficients in the differential operator. This means this lemma holds true also on meshes with $\mu < \lambda$ and S_h and S_h^* being the solution operator for the discretized interface problem. Consequently, Assumptions VAR1 and PP1 hold true.

Variational discrete approach We can directly formulate the error estimates for the variational discrete approach introduced in Subsection 5.1.2.1.

Theorem 5.46. *Let $\bar{u}$ be the solution of the optimal control problem (5.33) with the state equation (4.79) and $\bar{u}_h^s$ the solution of the corresponding variational discrete problem (5.8) on a mesh of type (2.12) with grading parameter $\mu < \lambda$. With the associated states $\bar{y} = S\bar{u}$, $\bar{y}_h^s = S_h \bar{u}_h^s$ and adjoint states $\bar{p} = P\bar{u}$, $\bar{p}_h^s = P_h \bar{u}_h^s$ the estimates*

$$\|\bar{u} - \bar{u}_h^s\|_{L^2(\Omega)} \leq ch^2 \left(\|\bar{u}\|_{L^2(\Omega)} + \|y_d\|_{L^2(\Omega)} \right),$$
$$\|\bar{y} - \bar{y}_h^s\|_{L^2(\Omega)} \leq ch^2 \left(\|\bar{u}\|_{L^2(\Omega)} + \|y_d\|_{L^2(\Omega)} \right),$$
$$\|\bar{p} - \bar{p}_h^s\|_{L^2(\Omega)} \leq ch^2 \left(\|\bar{u}\|_{L^2(\Omega)} + \|y_d\|_{L^2(\Omega)} \right)$$

hold with a constant c independent of h.

Proof. We have verified Assumption VAR1 above. The finite element error estimates of Assumption VAR2 are proved in Theorem 4.29 taking into account that $S_h = S_h^*$. Then Theorem 5.7 yields the assertion. $\qquad\square$

Postprocessing approach For the postprocessing approach of Subsection 5.1.2.2 it remains to check Assumptions PP3–PP4 since Assumption PP2 is already proved in Theorem 4.29.

Indeed, the proofs of Assumptions PP3 and PP4 are very similar to the case treated in Subsection 5.2.2. Nevertheless we sketch them here since one does not have to exploit anisotropic features as in the case of edge singularities such that the proofs may become clearer.

Before we state the next lemma, we should mention that Lemma 5.34 holds true in the two-dimensional setting, too [15, Lemma 5]. Furthermore Lemma 5.35 holds already for $p > 2$ since $L^\infty(\Omega) \hookrightarrow W^{1,p}(\Omega)$ for space dimension two.

As in Subsection 5.2.2 we use the sets

$$K_s = \bigcup_{\{T \in T_h : r_T = 0\}} T \quad \text{and} \quad K_r = \Omega \backslash \bar{K}_s.$$

Lemma 5.47. *The estimate*

$$\|Q_h \bar{p} - R_h \bar{p}\|_{L^2(\Omega)} \le ch^2 \left(\|\bar{u}\|_{L^\infty(\Omega} + \|y_d\|_{L^\infty(\Omega)} \right)$$

is true on a mesh of type (2.12) with grading parameter $\mu < \lambda$.

Proof. We split

$$\|Q_h \bar{p} + R_h \bar{p}\|_{L^2(\Omega)}^2 = \sum_{T \subset K_r} \|Q_h \bar{p} + R_h \bar{p}\|_{L^2(T)}^2 + \sum_{T \subset K_s} \|Q_h \bar{p} + R_h \bar{p}\|_{L^2(T)}^2. \tag{5.95}$$

For $T \in K_r$ we can write with the use of [15, Lemma 5]

$$\|Q_h \bar{p} + R_h \bar{p}\|_{L^2(T)}^2 = \frac{1}{|T|} \left| \int_T (\bar{p} - R_h \bar{p}) \, dx \right|^2 \le ch^4 |\bar{p}|_{V_{2-2\mu}^{2,2}(T)}^2.$$

This means

$$\|Q_h \bar{p} + R_h \bar{p}\|_{L^2(K_r)}^2 \le ch^2 \left(\sum_{T \subset K_r} |\bar{p}|_{V_{2-2\mu}^{2,2}(T)}^2 \right)^{1/2} \le ch^2 \|\bar{p}\|_{V_{2-2\mu}^{2,2}(\Omega)}. \tag{5.96}$$

In the following we choose $p > 2$, $p \ge 1/(1-\mu)$ and $p < 2/(1-\lambda)$. This is possible since $\lambda > 0$. Now we can apply Lemma 5.35 and get

$$\sum_{T \subset K_s} \|Q_h \bar{p} - R_h \bar{p}\|_{L^2(T)}^2 \le \sum_{T \subset K_s} |T|^{1-2/p} h_T^2 |\bar{p}|_{W^{1,p}(T)}^2$$

$$\le c \sum_{T \subset K_s} h_T^{4-4/p} |\bar{p}|_{W^{1,p}(T)}^2$$

$$\le c \left(\sum_{T \subset K_s} h_T^{(4-4/p)\frac{p-2}{p}} \right)^{\frac{p}{p-2}} \left(\sum_{T \subset K_s} |\bar{p}|_{W^{1,p}(T)}^p \right)^{2/p}$$

where we have used the Hölder inequality in the last step. Since K_s contains only a finite number of elements we can conclude from (2.12) and $1 - 1/p \geq \mu$

$$\|Q_h \bar{p} - R_h \bar{p}\|_{L^2(K_s)}^2 \leq ch^{\frac{4-4/p}{\mu}} \left(\sum_{T \in K_s} |\bar{p}|_{W^{1,p}(T)}^p \right)^{2/p} \leq ch^4 |\bar{p}|_{\mathcal{W}^{1,p}(\Omega)}^2.$$

This yields together with (5.95) and (5.96) and the a priori estimates of Lemma 5.44 and Corollary 5.45 the assertion. □

As for the proof of Lemma 5.40 we split the domain Ω into two parts,

$$K_1 := \bigcup_{T \in T_h : \bar{u} \notin V_{2-2\mu}^{2,2}(T)} T, \quad K_2 := \bigcup_{T \in T_h : \bar{u} \in V_{2-2\mu}^{2,2}(T)} T.$$

We assume again (comp. (5.85))

$$|K_1| \leq ch. \tag{5.97}$$

Lemma 5.48. *The estimate*

$$(Q_h \bar{u} - R_h \bar{u}, v_h)_{L^2(\Omega)} \leq ch^2 \|v_h\|_{L^\infty(\Omega)} \left(\|\bar{u}\|_{L^\infty(\Omega} + \|y_d\|_{L^\infty(\Omega)} \right) \qquad \forall v_h \in V_{0h}$$

holds on a mesh of type (2.12) *with grading parameter* $\mu < \lambda$.

Proof. We introduce the domains $K_{1,r} = K_1 \backslash \bar{K}_s$ and $K_{2,r} = K_2 \backslash \bar{K}_s$ and do the same splitting as in the proof of Lemma 5.40, compare (5.86). With Lemma 5.34 and (2.12) one can conclude

$$\sum_{T \subset K_{2,r}} \|v_h\|_{L^\infty(T)} \int_T (\bar{u} - R_h \bar{u}) \, dx \leq c \sum_{T \subset K_{2,r}} \|v_h\|_{L^\infty(T)} h_T^2 |\bar{u}|_{H^2(T)}$$

$$\leq ch^2 \|v_h\|_{L^\infty(\Omega)} \sum_{T \subset K_{2,r}} |\bar{u}|_{V_{2-2\mu}^{2,2}(T)} \tag{5.98}$$

$$\leq ch^2 \|v_h\|_{L^\infty(\Omega)} |K_{2,r}|^{1/2} |\bar{u}|_{V_{2-2\mu}^{2,2}(\Omega)}. \tag{5.99}$$

With the use of Lemma 5.34 and (5.97) it follows

$$\sum_{T \subset K_{1,r}} \|v_h\|_{L^\infty(T)} \int_T (\bar{u} - R_h \bar{u}) \, dx \leq \sum_{T \subset K_{1,r}} \|v_h\|_{L^\infty(T)} |T| h_T |\bar{u}|_{W^{1,\infty}(T)}$$

$$\leq ch^2 \|v_h\|_{L^\infty(\Omega)} |r^{1-\mu} \bar{u}|_{\mathcal{W}^{1,\infty}(\Omega)}. \tag{5.100}$$

Finally we estimate for the elements at the corner

$$\sum_{T \subset K_s} \|v_h\|_{L^\infty(T)} \int_T (\bar{u} - R_h \bar{u}) \, dx \leq c \sum_{T \subset K_s} |T| \|v_h\|_{L^\infty(T)} \|\bar{u}\|_{L^\infty(T)}$$

$$\leq ch^2 \|v_h\|_{L^\infty(\Omega)} \|\bar{u}\|_{L^\infty(\Omega)}, \tag{5.101}$$

where we have used the fact that K_s contains only a finite number of elements independent of h. Since $\bar{u}$ is constant on the active parts and equal to $-\bar{p}/\nu$ on the inactive part, we can substitute $\bar{u}$ with $\bar{p}$ in the seminorms in inequalities (5.99) and (5.100). The a priori estimates in Lemma 5.44 and Corollary 5.45 yield then together with (5.99)–(5.101) and the splitting (5.86) the assertion. □

We summarize the results in the following theorem.

Theorem 5.49. *Let $\bar{u}$ be the solution of the optimal control problem (5.33) with the state equation (4.79) and $\bar{u}_h$ the corresponding discrete solution of (5.12) on a mesh of type (2.12) with $\mu < \lambda$. Furthermore, let $\bar{y} = S\bar{u}$, $\bar{p} = P\bar{u}$, $\bar{y}_h = S_h\bar{u}_h$, $\bar{p}_h = S_h\bar{u}_h$ be the associated states and adjoint states and $\tilde{u}_h$ be the postprocessed control constructed by (5.14). Then the estimates*

$$\|\bar{y} - \bar{y}_h\|_{L^2(\Omega)} \leq ch^2 \left(\|\bar{u}\|_{C^{0,\sigma}(\bar{\Omega})} + \|y_d\|_{C^{0,\sigma}(\bar{\Omega})} \right),$$

$$\|\bar{p} - \bar{p}_h\|_{L^2(\Omega)} \leq ch^2 \left(\|\bar{u}\|_{C^{0,\sigma}(\bar{\Omega})} + \|y_d\|_{C^{0,\sigma}(\bar{\Omega})} \right),$$

$$\|\bar{u} - \tilde{u}_h\|_{L^2(\Omega)} \leq ch^2 \left(\|\bar{u}\|_{C^{0,\sigma}(\bar{\Omega})} + \|y_d\|_{C^{0,\sigma}(\bar{\Omega})} \right)$$

hold true.

Proof. The assertion follows from Theorem 5.14. Assumption PP1 is verified on page 123, Assumption PP2 in Theorem 4.29, Assumption PP3 in Lemma 5.47 and Assumption PP4 in Lemma 5.48. □

5.2.3.3 Numerical test

We consider the optimal control problem (5.33) with the state equation (4.79),

$$-k_i \Delta y_i = u_i + f_i \qquad \text{in } \Omega_i, \quad i = 1, 2, 3,$$
$$y_i(r, \omega_i) = y_{i+1}(r, \omega_i) \qquad i = 1, 2,$$
$$k_i \frac{\partial y_i(r, \omega_i)}{\partial \varphi} = k_{i+1} \frac{\partial y_{i+1}(r, \omega_i)}{\partial \varphi} \qquad i = 1, 2,$$
$$y = 0 \qquad \text{on } \partial\Omega,$$

and set

$$\Omega_i = \left\{ (r \cos\varphi, r \sin\varphi) \in \mathbb{R}^2 : 0 < r < 1, \ (i-1)\frac{\pi}{5} < \varphi < i\frac{\pi}{5} \right\} \text{ for } i = 1, 2, 3,$$
$$k_1 = k_3 = 1, \ k_2 = 50,$$
$$\omega_1 = \omega_2 = \omega_3 = \frac{\pi}{5}.$$

In [100, Example 2.29] it is shown that the smallest positive solution λ of (4.82) fulfills a trigonometrical equation. We used this fact to compute λ in our case as $\lambda \approx 0.31569$. The

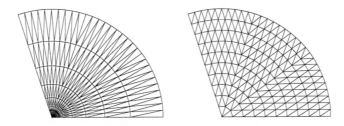

Figure 5.3: Graded mesh with $\mu = 0.3$ and quasi-uniform mesh ($\mu = 1$).

Table 5.7: $L^2(\Omega)$-errors of the computed control $\tilde{u}_h$, state $\bar{y}_h$ and adjoint state $\bar{p}_h$ on graded meshes ($\mu = 0.3$)

ndof	$\|u - \tilde{u}_h\|$	eoc	$\|y - \bar{y}_h\|$	eoc	$\|p - \bar{p}_h\|$	eoc
126	1.85e−01		2.40e−02		2.39e−04	
551	4.96e−02	1.78	6.38e−03	1.80	7.42e−05	1.58
2301	1.23e−02	1.95	1.63e−03	1.91	1.99e−05	1.84
6567	4.39e−03	1.97	5.85e−04	1.96	7.19e−06	1.94
23126	1.26e−03	1.98	1.68e−04	1.98	2.08e−06	1.97
93126	3.15e−04	1.99	4.21e−05	1.99	5.20e−07	1.99
373751	7.87e−05	1.99	1.05e−05	2.00	1.30e−07	1.99
1497501	1.96e−05	2.00	2.63e−06	2.00	3.25e−08	2.00

data f and y_d are chosen such that

$$\bar{y}_i = \frac{1}{\nu}\bar{p}_i = \frac{1}{k_i}(r^\lambda - r^{5/2})\sin\left(\frac{\pi}{\omega_i}\varphi\right),$$

$$\bar{u}_i = \Pi_{[-0.3,10]}\left(-\frac{1}{\nu}\bar{p}_i\right)$$

is the exact solution where $\nu = 10^{-3}$. On graded meshes with $\mu = 0.3 < \lambda$ one gets the predicted convergence rate of 2 for the L^2-error in control, state and adjoint state, see Table 5.7. On quasi-uniform meshes ($\mu = 1$) one observes a reduced convergence rate of approximately $1 + \lambda$, see Table 5.8. An example of the corresponding meshes is given in Figure 5.3.

5.3 Stokes equations as state equation

In this section we consider the optimal control problem (5.1) with the Stokes equations as state equations. The notation in Section 5.1 is motivated by the standard notation for the case of a scalar elliptic state equation. In the Stokes equations there occur velocity and

Table 5.8: $L^2(\Omega)$-errors of the computed control $\tilde{u}_h$, state $\bar{y}_h$ and adjoint state $\bar{p}_h$ on quasi-uniform meshes ($\mu = 1.0$)

ndof	$\|u - \tilde{u}_h\|$	eoc	$\|y - \bar{y}_h\|$	eoc	$\|p - \bar{p}_h\|$	eoc
126	5.28e−02		3.52e−02		8.79e−05	
551	1.63e−02	1.59	1.42e−02	1.23	2.51e−05	1.70
2301	6.13e−03	1.37	5.71e−03	1.27	7.78e−06	1.64
6567	3.03e−03	1.34	2.90e−03	1.29	3.47e−06	1.54
23126	1.30e−03	1.35	1.28e−03	1.30	1.39e−06	1.45
93126	5.17e−04	1.32	5.13e−04	1.31	5.31e−07	1.38
373751	2.07e−04	1.32	2.06e−04	1.31	2.09e−07	1.34
1497501	8.28e−05	1.32	8.27e−05	1.31	8.32e−08	1.33

pressure, such that the state in the optimal control problem has actually two components. As we may distinguish between velocity and pressure we slightly change the notation compared to Section 5.1. As in Section 4.4 about the finite element error analysis of the Stokes equations we denote by v the velocity field and by q the pressure. The velocity field v plays the role of the state y in Section 5.1. Consequently, we substitute y_d by v_d, such that the optimal control problem (5.1) reads as

$$J(\bar{u}) = \min_{u \in U^{\mathrm{ad}}} J(u)$$

$$J(u) := \frac{1}{2}\|Su - v_d\|^2_{L^2(\Omega)^d} + \frac{\nu}{2}\|u\|^2_{L^2(\Omega)}. \tag{5.102}$$

Here, S is the solution operator of the Stokes equations

$$\begin{aligned} -\Delta v - \nabla q &= u &&\text{in } \Omega, \\ \nabla \cdot v &= 0 &&\text{in } \Omega, \\ v &= 0 &&\text{on } \partial\Omega, \end{aligned}$$

and maps the control u to the velocity v. As in the scalar elliptic case, we assume the desired velocity field v_d to be Hölder continuous, i.e. $v_d \in C^{0,\sigma}(\bar{\Omega})^d$, $\sigma \in (0,1)$. As for the adjoint state the same notational problems occur as for the state, we introduce w and r as adjoint velocity field and adjoint pressure, respectively. Then the adjoint problem reads as, see, e.g., [112],

$$\begin{aligned} -\Delta w - \nabla r &= v - v_d &&\text{in } \Omega, \\ \nabla \cdot w &= 0 &&\text{in } \Omega, \\ w &= 0 &&\text{on } \partial\Omega. \end{aligned}$$

and its corresponding weak formulation

$$\text{Find } (w,r) \in X \times M:$$
$$\begin{aligned} a(\varphi, w) - b(\varphi, r) &= (f, \varphi) &&\forall \varphi \in X \\ b(v, \psi) &= 0 &&\forall \psi \in M \end{aligned}$$

with bilinear forms a and b as defined in Section 4.4. As in case of a scalar elliptic state equation we denote by S^* the solution operator of the adjoint problem, that means $w = S^*(v - v_d)$. We introduce also the affine operator P via $w = S^*(Su - v_d) = Pu$. Taking the modified notation into account the assertions of Theorem 5.4 and Lemma 5.6 can be reformulated like in the following lemma. Notice that in our case $Z = U = L^2(\Omega)^d$.

Lemma 5.50. *The optimal control problem (5.102) has a unique solution $\bar{u}$. The variational inequality*

$$(\bar{w} + \nu\bar{u}, u - \bar{u})_U \geq 0 \qquad \forall u \in U^{\text{ad}} \tag{5.103}$$

is a necessary and sufficient condition for the optimality of $\bar{u}$. The projection formula

$$\bar{u} = \Pi_{U^{\text{ad}}}\left(-\frac{1}{\nu}\bar{w}\right)$$

is an equivalent formulation for condition (5.103). Here, $\bar{w} = P\bar{u}$ is the corresponding adjoint velocity.

5.3.1 Prismatic domain

In this subsection we derive L^2-error estimates on anisotropic meshes for the optimal control problem (5.102) with the state equation (4.127), i.e., we consider the Stokes equations as state equation in the prismatic domain $\Omega = G \times Z \subset \mathbb{R}^3$ with a bounded polygonal domain $G \subset R^2$ and a interval $Z := (0, z_0) \subset \mathbb{R}$. We assume that G has only one corner with interior angle $\omega > \pi$ at the origin. The following results are originally published in [102].

5.3.1.1 Regularity

Lemma 5.51. *Let $\bar{u}$ be the solution of the optimal control problem (5.102) with the state equation (4.127). For the associated velocity field $\bar{v} = S\bar{u}$ and adjoint velocity field $\bar{w} = S\bar{u}$ one has $\bar{u} \in C^{0,\sigma}(\bar{\Omega})^3$, $\bar{v} \in C^{0,\sigma}(\bar{\Omega})^3$ and $\bar{w} \in C^{0,\sigma}(\bar{\Omega})^3$ for some $\sigma \in (0, 1/2)$. The a priori estimates*

$$\|\bar{v}\|_{C^{0,\sigma}(\bar{\Omega})^3} \leq c\|\bar{u}\|_{L^\infty(\Omega)} \leq c\|\bar{u}\|_{C^{0,\sigma}(\bar{\Omega})^3}, \tag{5.104}$$

$$\|\bar{w}\|_{C^{0,\sigma}(\bar{\Omega})^3} \leq c\left(\|\bar{u}\|_{L^\infty(\Omega)^3} + \|y_d\|_{C^{0,\sigma}(\bar{\Omega})^3}\right) \tag{5.105}$$

are valid.

Proof. For a value $\mu < \lambda$ it is $1 - \mu > 1 - \lambda$. From the fact that $\bar{u} \in L^2(\Omega)$ this yields with Lemma 4.36 that $\bar{v} \in V_{1-\mu}^{2,2}(\Omega)^3$. Since $\lambda > 1/2$ (see Remark 4.37) one can always choose a value for μ such that $1/2 < \mu < \lambda$. Then the embedding $V_{1-\mu}^{2,2}(\Omega) \hookrightarrow V_0^{2-(1-\mu),2}(\Omega) \hookrightarrow W^{1+\mu,2}(\Omega) \hookrightarrow L^\infty(\Omega)$ holds according to [113, Lemma 1.2], the Sobolev embedding theorem and the fact that $1 + \mu - 3/2 > 0$. This yields $\bar{v} \in L^\infty(\Omega)^3$ and therefore

$\bar{v} - v_{\mathrm{d}} \in L^{\infty}(\Omega)^3$. Applying Lemma 4.36 to the adjoint equation yields $\bar{w} \in V_{\beta}^{2,p}(\Omega)^3$ for all $p > 1$ and $\beta > 2 - \lambda - 2/p$. In the following we choose β such that

$$2 - \lambda - \frac{2}{p} < \beta < 2 - \frac{3}{p} - \sigma.$$

This is possible as long as $\lambda > 1/p + \sigma$. Since $\sigma < 1/2$ and $\lambda > 1/2$ this can be guaranteed for p large enough. With this setting the embedding

$$V_{\beta}^{2,p}(\Omega) \hookrightarrow V_0^{2-\beta,p}(\Omega) \hookrightarrow W^{2-\beta,p}(\Omega) \hookrightarrow C^{0,\sigma}(\bar{\Omega}) \tag{5.106}$$

holds, where we have utilized [113, Lemma 1.2] and Sobolev's embedding theorem again. It follows $\bar{w} \in C^{0,\sigma}(\bar{\Omega})^3$. The projection formula (5.6) yields $\bar{u} \in C^{0,\sigma}(\bar{\Omega})^3$. With the same argumentation for the state equation one can conclude $\bar{v} \in C^{0,\sigma}(\bar{\Omega})^3$. The estimate (5.104) follows then from Lemma 4.36,

$$\|\bar{v}\|_{C^{0,\sigma}(\bar{\Omega})^3} \leq c\|\bar{v}\|_{V_{\beta}^{2,p}(\Omega)^3} \leq c\|\bar{u}\|_{L^p(\Omega)^3} \leq c\|\bar{u}\|_{L^{\infty}(\Omega)^3} \leq c\|\bar{u}\|_{C^{0,\sigma}(\bar{\Omega})^3}$$

where we have chosen p large enough. For the proof of the inequality (5.105) we conclude with the use of Lemma 4.36 and the embedding (5.106)

$$\|\bar{w}\|_{C^{0,\sigma}(\bar{\Omega})^3} \leq c\|\bar{w}\|_{V_{\beta}^{2,p}(\Omega)^3} \leq c\|\bar{v} - v_{\mathrm{d}}\|_{L^p(\Omega)^3} \leq c\left(\|\bar{v}\|_{L^{\infty}(\Omega)^3} + \|v_{\mathrm{d}}\|_{L^{\infty}(\Omega)^3}\right).$$

In the last step we have used the embedding $L^{\infty}(\Omega) \hookrightarrow L^p(\Omega)$ and the triangle inequality. The embedding $C^{0,\sigma}(\bar{\Omega}) \hookrightarrow L^{\infty}(\Omega)$ and the application of estimate (5.104) yield finally inequality (5.105). $\qquad\square$

In the proof of the foregoing lemma we have implicitly also shown the regularity of $\bar{v}$ and $\bar{w}$ in weighted Sobolev spaces. We collect these results in the following corollary.

Corollary 5.52. *Let $\bar{u}$ be the solution of the optimal control problem (5.102) with the state equation (4.127). For the associated velocity field $\bar{v} = S\bar{u}$ and adjoint velocity field $\bar{w} = S\bar{u}$ one has $\bar{v} \in V_{\beta}^{2,p}(\Omega)^3$, $\bar{w} \in V_{\beta}^{2,p}(\Omega)^3$ for $\beta > 2 - \lambda - 2/p$. The a priori estimates*

$$\|\bar{v}\|_{V_{\beta}^{2,p}(\Omega)^3} \leq c\|\bar{u}\|_{L^{\infty}(\Omega)^3} \leq c\|\bar{u}\|_{C^{0,\sigma}(\bar{\Omega})^3}, \tag{5.107}$$

$$\|\bar{w}\|_{V_{\beta}^{2,p}(\Omega)^3} \leq c\left(\|\bar{u}\|_{L^{\infty}(\Omega)^3} + \|v_{\mathrm{d}}\|_{L^{\infty}(\Omega)^3}\right) \tag{5.108}$$

hold for some $\sigma \in (0, 1/2)$.

Corollary 5.53. *Let $\bar{w}$ be the optimal adjoint velocity for the optimal control problem (5.102) with the state equation (4.127). Then one has $\bar{w} \in W^{1,p}(\Omega)^3$ and*

$$\|\bar{w}\|_{W^{1,p}(\Omega)^3} \leq c\left(\|\bar{u}\|_{L^{\infty}(\Omega)^3} + \|v_{\mathrm{d}}\|_{L^{\infty}(\Omega)^3}\right) \tag{5.109}$$

with $p < \frac{2}{1-\lambda}$. Furthermore it is $\bar{w} \in V_{1-\mu}^{1,p}(\Omega)$ and

$$\|\bar{w}\|_{V_{1-\mu}^{1,p}(\Omega)^3} \leq c\left(\|\bar{u}\|_{L^{\infty}(\Omega)^3} + \|v_{\mathrm{d}}\|_{L^{\infty}(\Omega)^3}\right) \tag{5.110}$$

for all $p > 1$ and $\mu < \lambda + \frac{2}{p}$.

Proof. Since $p < \frac{2}{1-\lambda}$ it is $1 > 2 - \lambda - \frac{2}{p}$. Therefore we can choose $\beta = 1$ in Corollary 5.52. Since $V_1^{2,p}(\Omega)^3 \hookrightarrow V_0^{1,p}(\Omega)^3 \hookrightarrow W^{1,p}(\Omega)^3$ the assertion (5.109) follows from inequality (5.108). For the proof of (5.110) we set $\beta = 2 - \mu$ in (5.108). This is possible due to the fact that $2 - \mu > 2 - \lambda - \frac{2}{p}$ since $\mu < \lambda + \frac{2}{p}$. The embedding $V_{2-\mu}^{2,p}(\Omega) \hookrightarrow V_{1-\mu}^{1,p}(\Omega)$ yields the assertion. $\qquad\square$

Lemma 5.54. *Let* $v_{\mathrm{d}} \in C^{0,\sigma}(\bar{\Omega})$, $\sigma \in (0, 1/2)$, *and* $\gamma > 1 - \lambda$. *Then the inequality*

$$\|r^\gamma \nabla P\bar{u}\|_{L^\infty(\Omega)^3} \le c \left(\|\bar{u}\|_{C^{0,\sigma}(\bar{\Omega})^3} + \|v_{\mathrm{d}}\|_{C^{0,\sigma}(\bar{\Omega})^3} \right)$$

is valid.

Proof. In order to prove the assertion we utilize Theorem 6.1 of [92]. We set $l = 2$ and $\delta = \beta$ in that Theorem. This results in the condition $2 - \lambda < \delta - \sigma < 2$, what is equivalent to

$$1 - \lambda < \delta - \sigma - 1 < 1. \tag{5.111}$$

Since $\bar{v} - v_{\mathrm{d}} = S\bar{u} - v_{\mathrm{d}} \in C^{0,\sigma}(\bar{\Omega})$ (comp. Lemma 5.51) we can conclude $P\bar{u} \in C_{\delta,\delta}^{2,\sigma}(\Omega)^3$ for δ satisfying (5.111). The definition of this weighted Hölder space is given on page 1013 of [92]. Taking this definition into account, one can conclude

$$r^{\delta-1-\sigma} \nabla P\bar{u} \in L^\infty(\Omega)^3.$$

If we set $\gamma = \delta - \sigma - 1$, it finally follows

$$\|r^\gamma \nabla P\bar{u}\|_{L^\infty(\Omega)^3} \le c\|\bar{v} - v_{\mathrm{d}}\|_{C^{0,\sigma}(\Omega)^3} \text{ for } \gamma > 1 - \lambda.$$

The application of the triangle inequality and inequality (5.104) yield the assertion. $\qquad\square$

5.3.1.2 Approximation error estimate in $L^2(\Omega)$

As in Subsection 4.4.3 we consider an anisotropic triangulation of Ω following (2.13). The state and adjoint state equation are discretized by a non-conforming finite element scheme, namely the lower order Crouzeix-Raviart finite element space X_h,

$$X_h := \left\{ v_h \in L^2(\Omega)^3 : v_h|_T \in (\mathcal{P}_1)^3 \; \forall T, \int_F [v_h]_F = 0 \; \forall F \right\}$$

for the velocity and the space of piecewise constant functions M_h,

$$M_h := \left\{ q_h \in L^2(\Omega) : q_h|_T \in \mathcal{P}_0 \; \forall T, \int_\Omega q_h = 0 \right\}$$

for the pressure. The solution mappings S_h and S_h^p of the discretized state equation are defined such that one has for all $(\varphi_h, \psi_h) \in X_h \times M_h$ and $u \in U$

$$a_h(S_h u, \varphi_h) + b_h(\varphi_h, S_h^p u) = (u, \varphi_h) \quad \text{and} \quad b_h(S_h u, \psi_h) = 0.$$

with a_h, b_h as in (4.87). Analogously, we introduce S_h^* and $S_h^{p,*}$ as solution mappings of the discrete adjoint equation and the operator P_h such that $P_h u = S_h^*(S_h u - v_{\mathrm{d}}) = w_h$.

In the remainder of this subsection we will verify Assumptions VAR1 and VAR2 and Assumptions PP1–PP4 and thus prove the error estimates stated in Theorem 5.7 and Theorem 5.14, respectively. In a first step we check Assumptions VAR1 and PP1. To this end we recall a discrete Poincaré inequality as it is proved in [82].

Lemma 5.55. *[82, Corollary 5.4] The discrete Poincaré inequality*

$$\|v_h\|_{L^2(\Omega)^d} \le c\|v_h\|_{X_h} \qquad \forall v_h \in X_h$$

holds.

With this lemma we can prove the boundedness of S_h and S_h^* which is not obvious because of the anisotropic discretization.

Lemma 5.56. *The discrete solution operators S_h and S_h^* are bounded,*

$$\|S_h\|_{U \to U} \le c, \qquad\qquad \|S_h^*\|_{U \to U} \le c,$$
$$\|S_h\|_{U \to X_h} \le c, \qquad\qquad \|S_h^*\|_{U \to X_h} \le c,$$
$$\|S_h\|_{U \to L^\infty(\Omega)^d} \le c, \qquad \|S_h^*\|_{U \to L^\infty(\Omega)^d} \le c$$

with constants c independent of h.

Proof. We show this lemma for the operator S_h, the proofs for S_h^* are analogous. The first estimate follows with

$$\|S_h u\|_U \le \|Su\|_U + \|S_h u - Su\|_U$$

from the boundedness of S as operator from U to U and inequality (4.92). The subtraction of the equations (4.88) and (4.89) with $v_h = S_h u$ yields

$$a_h(S_h u, \varphi_h) + b_h(\varphi_h, q_h) - b_h(S_h u, \psi_h) = (u, \varphi_h) \quad \forall(\varphi_h, \psi_h) \in X_h \times M_h.$$

If one chooses $(\varphi_h, \psi_h) = (S_h u, q_h)$ this implies

$$a_h(S_h u, S_h u) = (u, S_h u).$$

Therefore we can estimate

$$\|S_h u\|_{X_h}^2 = a_h(S_h u, S_h u) = (u, S_h u)$$
$$\le c\|u\|_U\|S_h u\|_U \le c\|u\|_U\|S_h u\|_{X_h},$$

where we have used the Cauchy-Schwarz inequality and the discrete Poincaré inequality from Lemma 5.55. Division by $\|S_h u\|_{X_h}$ yields $\|S_h u\|_{X_h} \le c\|u\|_U$ and the second estimate is proved. The third estimate follows from the boundedness of S and inequality (4.94),

$$\|S_h u\|_{L^\infty(\Omega)^d} \le \|Su - S_h u\|_{L^\infty(\Omega)^d} + \|Su\|_{L^\infty(\Omega)^d} \le c\|u\|_U.$$

$\square$

Variational discrete approach The discrete control $\bar{u}_h^s$ is defined via (comp. (5.8))

$$J_h(\bar{u}_h^s) = \min_{u \in U^{\mathrm{ad}}} J_h(u),$$

$$J_h(u) := \frac{1}{2} \|S_h u - v_{\mathrm{d}}\|_{L^2(\Omega)}^2 + \frac{\nu}{2} \|u\|_{L^2(\Omega)}^2. \tag{5.112}$$

The boundedness of the operators S_h and S_h^* is proved in Lemma 5.56 such that Assumption VAR1 is guaranteed. Assumption VAR2 follows from the finite element error estimates in Lemma 4.39. Therefore we can summarize the error estimates that are given in Theorem 5.7 in the following theorem.

Theorem 5.57. *Let $\bar{u}$ be the solution of the optimal control problem (5.102) with the state equation (4.127) and $\bar{u}_h^s$ the corresponding variational discrete solution of (5.112) on a mesh of type (2.13) with $\mu < \lambda$. Furthermore, let $\bar{v} = S u$, $\bar{w} = P\bar{u}$, $\bar{v}_h^s = S_h \bar{u}_h^s$, $\bar{w}_h^s = S_h \bar{u}_h^s$ be the associated velocity and adjoint velocity. Then the estimates*

$$\|\bar{v} - \bar{v}_h^s\|_U \le ch^2 \left(\|\bar{u}\|_{C^{0,\sigma}(\bar{\Omega})^d} + \|v_{\mathrm{d}}\|_{C^{0,\sigma}(\bar{\Omega})^d} \right),$$

$$\|\bar{w} - \bar{w}_h^s\|_U \le ch^2 \left(\|\bar{u}\|_{C^{0,\sigma}(\bar{\Omega})^d} + \|v_{\mathrm{d}}\|_{C^{0,\sigma}(\bar{\Omega})^d} \right),$$

$$\|\bar{u} - \tilde{u}_h^s\|_U \le ch^2 \left(\|\bar{u}\|_{C^{0,\sigma}(\bar{\Omega})^d} + \|v_{\mathrm{d}}\|_{C^{0,\sigma}(\bar{\Omega})^d} \right)$$

are valid with a positive constant c independent of h.

Postprocessing approach For this approach we consider the discretized optimal control problem

$$J_h(\bar{u}_h) = \min_{u_h \in U_h^{\mathrm{ad}}} J_h(u_h),$$

$$J_h(u_h) := \frac{1}{2} \|S_h u_h - v_{\mathrm{d}}\|_{L^2(\Omega)}^2 + \frac{\nu}{2} \|u_h\|_{L^2(\Omega)}^2 \tag{5.113}$$

and improve the approximation of $\bar{u}$ by a postprocessing step (comp. (5.14)),

$$\tilde{u}_h = \Pi_{[u_a, u_b]} \left(-\frac{1}{\nu} \bar{w}_h \right). \tag{5.114}$$

The Assumption PP1 is already proved in Lemma 5.56, Assumption PP2 follows from Lemma 4.39.

It remains to check Assumptions PP3 and PP4. The regularity of the adjoint state plays a crucial role in that proofs. Since the regularity properties of each component of the velocity field of the Stokes problem are similar to those of the solution of the Poisson equation one can think of a componentwise consideration of the arguments in the proofs of Lemma 5.39 and Lemma 5.40. The drawback is that the results concerning the regularity of the solution along the edge in the space $L^p(\Omega)$ for general p (comp. Lemma 4.17) are not available for the Stokes problem. There we have only results for $p = 2$ (comp. Lemma 4.36). For this reason the proofs has to be modified for the Stokes problem.

Lemma 5.58. *Let the mesh be graded according to (2.13) with $\mu < \lambda$. Then the inequality*

$$\|Q_h \bar{w} - R_h \bar{w}\|_{L^2(\Omega)^3} \leq ch^2 \left(\|\bar{u}\|_{C^{0,\sigma}(\bar{\Omega})^3} + \|v_{\mathrm{d}}\|_{C^{0,\sigma}(\bar{\Omega})^3} \right)$$

holds.

Proof. We write

$$\|Q_h \bar{w} - R_h \bar{w}\|_{L^2(\Omega)^3}^2 = \|Q_h \bar{w} - R_h \bar{w}\|_{L^2(K_r)^3}^2 + \|Q_h \bar{w} - R_h \bar{w}\|_{L^2(K_s)^3}^2 \qquad (5.115)$$

with K_r and K_s as defined in (5.79). First we prove the estimate in K_r. Notice, that one has $\bar{w} \in H^2(K_r)^3$. We write for each component $\bar{w}_k$, $k = 1, 2, 3$, of $\bar{w} = (\bar{w}_1, \bar{w}_2, \bar{w}_3)$

$$\|Q_h \bar{w}_k - R_h \bar{w}_k\|_{L^2(K_r)}^2 = \sum_{T \subset K_r} \|Q_h \bar{w}_k - R_h \bar{w}_k\|_{L^2(T)}^2$$

$$= \sum_{T \subset K_r} |T|^{-1} \left| \int_T (\bar{w}_k - R_h \bar{w}_k) \, \mathrm{d}x \right|^2.$$

Now we can apply Lemma 5.34 and get

$$\|Q_h \bar{w}_k - R_h \bar{w}_k\|_{L^2(K_r)}^2 \leq \sum_{T \subset K_r} |T|^{-1} \left[c|T|^{1/2} \sum_{|\alpha|=2} h_T^\alpha \|D^\alpha \bar{w}_k\|_{L^2(T)} \right]^2$$

$$\leq c \sum_{T \subset K_r} \left[\sum_{|\alpha|=2} h_T^\alpha \|D^\alpha \bar{w}_k\|_{L^2(T)} \right]^2$$

$$\leq c \sum_{T \subset K_r} \left[ch^2 \left(\sum_{i=1}^2 \sum_{j=1}^2 \|r^{2-2\mu} \partial_{ij} \bar{w}_k\|_{L^2(T)} + \right. \right.$$

$$\left. \left. \sum_{i=1}^2 \|r^{1-\mu} \partial_{3i} \bar{w}_k\|_{L^2(T)} + \|\partial_{33} \bar{w}_k\|_{L^2(T)} \right) \right]^2$$

$$\leq ch^4 \left(|\bar{w}_k|_{V_{2-2\mu}^{2,2}(K_r)}^2 + |\partial_3 \bar{w}_k|_{V_{1-\mu}^{1,2}(K_r)}^2 + |\partial_{33} \bar{w}_k|_{V_0^{0,2}(K_r)}^2 \right).$$

This yields

$$\|Q_h \bar{w} - R_h \bar{w}\|_{L^2(K_r)^3} \leq ch^2 \left(|\bar{w}|_{V_{2-2\mu}^{2,2}(K_r)^3}^2 + |\partial_3 \bar{w}|_{V_{1-\mu}^{1,2}(K_r)^3}^2 + |\partial_{33} \bar{w}|_{V_0^{0,2}(K_r)^3}^2 \right)^{1/2}.$$

With the a priori estimates of Lemma 4.36, the embedding $C^{0,\sigma}(\bar{\Omega}) \hookrightarrow L^2(\Omega)$ and Lemma 5.51 one gets

$$\|Q_h \bar{w} - R_h \bar{w}\|_{L^2(K_r)^3} \leq ch^2 \|v - v_{\mathrm{d}}\|_{L^2(\Omega)} \leq ch^2 \left(\|\bar{u}\|_{C^{0,\sigma}(\bar{\Omega})^3} + \|\bar{v}_{\mathrm{d}}\|_{C^{0,\sigma}(\bar{\Omega})^3} \right). \quad (5.116)$$

We proceed with the estimate in the subdomain K_s. We choose p and γ such that

$$p > 3, \quad p < \frac{2}{1-\lambda}, \quad p < \frac{2}{\gamma}, \quad \gamma < 1 - \mu \text{ and } \gamma > 1 - \lambda. \tag{5.117}$$

Since $\lambda > \frac{1}{2}$ one has $3 < \frac{2}{1-\lambda}$. Further it is $3 < \frac{2}{\gamma}$ if $\gamma < \frac{2}{3}$. This can be fulfilled since $\frac{2}{3} > 1 - \lambda$. Finally $1 - \lambda < 1 - \mu$ due to the fact that $\mu < \lambda$. Altogether this means, that there are actually p and γ that satisfy the assumptions in (5.117). From Corollary 5.53 one has $\bar{w} \in W^{1,p}(\Omega)^3$. Now we can apply Lemma 5.35 on every component $\bar{w}_k$, $k = 1, 2, 3$, of $\bar{w}$ and conclude

$$\|Q_h \bar{w}_k - R_h \bar{w}_k\|^2_{L^2(K_s)} = \sum_{T \subset K_s} \|Q_h \bar{w}_k - R_h \bar{w}_k\|^2_{L^2(T)}$$

$$\leq c \sum_{T \in K_s} |T|^{1-2/p} \left(\sum_{|\alpha|=1} h_T^\alpha \|D^\alpha \bar{w}_k\|_{L^p(T)} \right)^2$$

$$\leq c \sum_{|\alpha|=1} \sum_{T \in K_s} |T|^{1-2/p} h_T^{2\alpha} \|D^\alpha \bar{w}_k\|^2_{L^p(T)}.$$

Since $h_T^{2\alpha} \leq ch^2$ for all $|\alpha| = 1$, one can continue with Lemma 5.54,

$$\|Q_h \bar{w}_k - R_h \bar{w}_k\|^2_{L^2(K_s)} \leq ch^2 \sum_{T \in K_s} |T|^{1-2/p} \|r^{-\gamma} r^\gamma \nabla \bar{w}_k\|^2_{L^p(T)}$$

$$\leq ch^2 \|r^\gamma \nabla \bar{w}_k\|^2_{L^\infty(\Omega)} \sum_{T \subset K_s} |T|^{1-2/p} \|r^{-\gamma}\|^2_{L^p(T)}. \tag{5.118}$$

In the following we prove that the inequality

$$\sum_{T \subset K_s} |T|^{1-2/p} \|r^{-\gamma}\|^2_{L^p(T)} \leq ch^2$$

is valid. To this end we apply the Hölder inequality and get

$$\sum_{T \subset K_s} |T|^{1-2/p} \|r^{-\gamma}\|^2_{L^p(T)} \leq \left[\sum_{T \subset K_s} \left(|T|^{1-2/p} \right)^{\frac{p}{p-2}} \right]^{\frac{p-2}{p}} \left[\sum_{T \subset K_s} \|r^{-\gamma}\|^p_{L^p(T)} \right]^{\frac{2}{p}}$$

$$\leq c \left(\sum_{T \subset K_s} |T| \right)^{\frac{p-2}{p}} \left(\int_0^{h^{1/\mu}} r^{-\gamma p} r \, dr \right)^{\frac{2}{p}}$$

$$\leq c |K_s|^{\frac{p-2}{p}} \left(h^{\frac{1}{\mu}(2-\gamma p)} \right)^{\frac{2}{p}}$$

where we have used $\gamma < \frac{2}{p}$ (comp. (5.117)) in the last step. Because $|K_s| \leq ch^{2/\mu}$ one can conclude

$$\sum_{T \subset K_s} |T|^{1-2/p} \|r^{-\gamma}\|^2_{L^p(T)} \leq ch^{\frac{2}{\mu}\left(\frac{p-2}{p} + \frac{2-\gamma p}{p}\right)} = ch^{\frac{2}{\mu}(1-\gamma)} \leq ch^2$$

135

using the fact that $\mu < 1 - \gamma$. This estimate yields together with (5.118) the inequality

$$\|Q_h\bar{w} - R_h\bar{w}\|_{L^2(K_s)^3} \le ch^2 \|r^\gamma \nabla \bar{w}\|_{L^\infty(\Omega)^3}$$

and with Lemma 5.54

$$\|Q_h\bar{w} - R_h\bar{w}\|_{L^2(K_s)^3} \le ch^2 \left(\|\bar{u}\|_{C^{0,\sigma}(\bar{\Omega})^3} + \|\bar{v}_d\|_{C^{0,\sigma}(\bar{\Omega})^3} \right).$$

Together with estimate (5.116) and equality (5.115) this yields the assertion. $\qquad \square$

The following lemma proves Assumption PP4. Due to the weaker regularity results for derivatives of the solution in edge direction, we need the slightly stronger condition

$$\#K_1 \le ch^{-2} \tag{5.119}$$

in comparison to (5.85) for the set K_1 defined in (5.84).

Lemma 5.59. *Let T_h be an anisotropic, graded mesh satisfying (2.13) with $\mu < \lambda$. Let $\bar{u}$ be the solution of the optimal control problem (5.102). Then the estimate*

$$(Q_h\bar{u} - R_h\bar{u}, \varphi_h)_{L^2(\Omega)} \le ch^2 \|\varphi_h\|_{L^\infty(\Omega)} \left(\|\bar{u}\|_{L^\infty(\Omega)} + \|v_d\|_{C^{0,\sigma}(\bar{\Omega})} \right)$$

is valid for all $\varphi_h \in X_h$ provided that the assumption (5.119) holds.

Proof. For the proof we refer to the proof of Lemma 5.40. If one substituted v_h by φ_h in the estimates (5.86) and (5.87), they are valid for every component of $\bar{u} = (u_1, u_2, u_3)$ and of $\varphi_h = (\varphi_{h,1}, \varphi_{h,2}, \varphi_{h,3})$. We only have to modify the estimate for $K_{1,r}$ since we do not have an estimate of type (5.74) for $\partial_3\bar{w}$. Therefore we write

$$\sum_{T \subset K_{1,r}} \|\varphi_{h,k}\|_{L^\infty(T)} |T| \sum_{|\alpha|=1} h_T^\alpha \|D^\alpha \bar{u}_k\|_{L^\infty(T)}$$

$$\le c\|\varphi_{h,k}\|_{L^\infty(\Omega)} \sum_{T \subset K_{1,r}} |T| \left(hr_T^{1-\mu} \sum_{i=1}^{2} \|\partial_i \bar{u}_k\|_{L^\infty(T)} + h\|\partial_3 \bar{u}_k\|_{L^\infty(T)} \right)$$

$$\le c\|\varphi_{h,k}\|_{L^\infty(\Omega)} h^4 \sum_{T \subset K_{1,r}} r_T^{2-2\mu} \left(\sum_{i=1}^{2} \|r^{1-\mu}\partial_i \bar{u}_k\|_{L^\infty(T)} + r^{\mu-1}\|r^{1-\mu}\partial_3 \bar{u}_k\|_{L^\infty(T)} \right)$$

$$\le ch^4 \#K_{1,r} \|\varphi_{h,k}\|_{L^\infty(\Omega)} \|r^{1-\mu}\nabla\bar{u}_k\|_{L^\infty(K_{1,r})}$$

$$\le ch^2 \|\varphi_{h,k}\|_{L^\infty(\Omega)} \|r^{1-\mu}\nabla\bar{u}_k\|_{L^\infty(\Omega)}.$$

where we utilized assumption (5.119) in the last step. Like in the proof of Lemma 5.40 we end up with

$$(\varphi_h, Q_h\bar{u} - R_h\bar{u}) \le \frac{c}{\nu} h^2 \|\varphi_h\|_{L^\infty(\Omega)^3}.$$

$$\left(\sum_{i=1}^{2}\sum_{j=1}^{2} \|r^{2-2\mu}\partial_{ij}\bar{w}\|_{L^2(K_{2,r})^3} + \sum_{i=1}^{2} \|r^{1-\mu}\partial_{3i}\bar{w}\|_{L^2(K_{2,r})^3} + \|\partial_{33}\bar{w}\|_{L^2(K_{2,r})^3} \right.$$

$$\left. + \sum_{i=1}^{2} \|r^{1-\mu}\partial_i\bar{w}\|_{L^\infty(K_{1,r})^3} + \|\partial_3\bar{w}\|_{L^\infty(K_{1,r})^3} + \nu\|\bar{u}\|_{L^\infty(K_s)^3} \right).$$

Finally, the application of Lemma 4.36 and Corollary 5.54 yields the assertion. □

The following theorem summarizes the discretization error estimates for the optimal control problem (5.102).

Theorem 5.60. *Let $\bar{u}$ be the solution of the optimal control problem (5.102) with the state equation (4.127) and $\bar{u}_h$ the corresponding discrete solution of (5.113) on a mesh of type (2.13) with $\mu < \lambda$. Furthermore, let $\bar{v} = S\bar{u}$, $\bar{w} = P\bar{u}$, $\bar{v}_h = S_h\bar{u}_h$, $\bar{w}_h = S_h\bar{u}_h$ be the associated states and adjoint states and $\tilde{u}_h$ be the postprocessed control constructed by (5.114). Then the estimates*

$$\|\bar{v} - \bar{v}_h\|_U \leq ch^2 \left(\|\bar{u}\|_{C^{0,\sigma}(\bar{\Omega})^d} + \|v_d\|_{C^{0,\sigma}(\bar{\Omega})^d} \right),$$

$$\|\bar{w} - \bar{w}_h\|_U \leq ch^2 \left(\|\bar{u}\|_{C^{0,\sigma}(\bar{\Omega})^d} + \|v_d\|_{C^{0,\sigma}(\bar{\Omega})^d} \right),$$

$$\|\bar{u} - \tilde{u}_h\|_U \leq ch^2 \left(\|\bar{u}\|_{C^{0,\sigma}(\bar{\Omega})^d} + \|v_d\|_{C^{0,\sigma}(\bar{\Omega})^d} \right)$$

are valid with a positive constant c independent of h.

Proof. The estimates follow from Theorem 5.14. Assumption PP1 is proved in Lemma 5.56, Assumption PP2 in Lemma 4.39, Assumption PP3 in Lemma 5.58 and Assumption PP4 in Lemma 5.59. □

5.3.1.3 Numerical test

We illustrate our theoretical findings by a numerical example. In order to be able to construct an analytical solution we consider the slightly modified functional

$$J(v, u) := \frac{1}{2}\|v - v_d\|^2_{L^2(\Omega)^d} + \frac{\nu}{2}\|u\|_{L^2(\Omega)^d} + \int_{\partial\Omega} \frac{\partial v}{\partial n} g \, ds,$$

and the state equation

$$\begin{aligned}
-\Delta v + \nabla q &= u + f &&\text{in } \Omega, \\
\nabla \cdot v &= 0 &&\text{in } \Omega, \\
v &= g &&\text{on } \partial\Omega.
\end{aligned}$$

The adjoint equation is given as

$$\begin{aligned}
-\Delta w - \nabla r &= v - v_d &&\text{in } \Omega, \\
\nabla \cdot w &= 0 &&\text{in } \Omega, \\
w &= g &&\text{on } \partial\Omega.
\end{aligned}$$

where the inhomogeneous boundary conditions are the result of the last integral term in the functional J. The domain Ω is set as

$$\Omega = \left\{ (r\cos\varphi, r\sin\varphi, x_3) \in \mathbb{R}^3 : 0 < r < 1, 0 < \varphi < \frac{3}{2}\pi, 0 < x_3 < 1 \right\}.$$

Table 5.9: $L^2(\Omega)$-errors of the computed control $\tilde{u}_h$, velocity $\bar{v}_h$ and adjoint velocity $\bar{w}_h$ on anisotropic graded meshes ($\mu = 0.4$)

ndof	$\|u - \tilde{u}_h\|$	eoc	$\|v - \bar{v}_h\|$	eoc	$\|w - \bar{w}_h\|$	eoc
12225	8.21e−03		1.36e−02		1.36e−03	
34251	4.61e−03	1.68	7.58e−03	1.71	7.63e−03	1.69
101400	2.45e−03	1.75	3.95e−03	1.80	3.97e−03	1.80
346275	1.16e−03	1.83	1.84e−03	1.87	1.84e−03	1.88
825600	6.69e−04	1.89	1.05e−03	1.92	1.06e−03	1.92
1618125	4.35e−04	1.92	6.80e−04	1.95	6.82e−04	1.95
2802600	3.06e−04	1.93	4.75e−04	1.96	4.77e−04	1.96
6662400	1.74e−04	1.95	2.69e−04	1.97	2.70e−04	1.97

The functions f, g and v_d are chosen such that

$$\bar{v} = \bar{w} = \begin{pmatrix} x_3(x_3 - 1)r^\lambda \Phi_1(\varphi) \\ x_3(x_3 - 1)r^\lambda \Phi_2(\varphi) \\ r^{2/3} \sin \frac{2}{3}\varphi \end{pmatrix}, \quad \bar{q} = -\bar{r} = x_3(x_3 - 1)r^{\lambda-1}\Phi_p(\varphi),$$

$$\bar{u} = \Pi_{[-2.0, 0.1]}\left(-\frac{1}{\nu}\bar{w}\right)$$

is the exact solution for the optimal control problem. Here, $\lambda \approx 0.5445$ is the smallest positive solution of the eigenvalue problem (4.128). The functions Φ_1, Φ_2 and Φ_p are given as

$$\Phi_1(\varphi) = -\sin(\lambda\varphi)\cos\omega - \lambda\sin(\varphi)\cos(\lambda(\omega - \varphi) + \varphi)$$
$$+ \lambda\sin(\omega - \phi)\cos(\lambda\varphi - \varphi) + \sin(\lambda(\omega - \varphi)),$$
$$\Phi_2(\varphi) = -\sin(\lambda\varphi)\sin\omega - \lambda\sin(\varphi)\sin(\lambda(\omega - \varphi) + \varphi)$$
$$- \lambda\sin(\omega - \varphi)\sin(\lambda\varphi - \varphi),$$
$$\Phi_p(\varphi) = 2\lambda\left[\sin((\lambda - 1)\varphi + \omega) + \sin((\lambda - 1)\varphi - \lambda\omega)\right].$$

This solution has the typical singular behavior near the edge (comp. [13]).

In Table 5.9 one can observe second order convergence in the post-processed control $\tilde{u}_h$ as well as in the approximated velocity $\bar{v}_h$ and adjoint velocity $\bar{w}_h$ for sufficiently graded meshes ($\mu = 0.4 < 0.5445 = \lambda$). This confirms our theoretical findings. On quasi-uniform meshes the convergence rates are significantly smaller, see Table 5.10.

5.3.2 Polygonal domain

In this subsection we comment on the optimal control problem (5.102) with the state equation (4.105).

Table 5.10: $L^2(\Omega)$-errors of the computed control $\tilde{u}_h$, velocity $\bar{v}_h$ and adjoint velocity $\bar{w}_h$ on quasi-uniform meshes ($\mu = 1.0$)

ndof	$\|u - \tilde{u}_h\|$	eoc	$\|v - \bar{v}_h\|$	eoc	$\|w - \bar{w}_h\|$	eoc
12225	8.99e−03		1.21e−02		1.21e−02	
34251	5.38e−03	1.49	6.98e−03	1.60	7.04e−03	1.58
101400	3.26e−03	1.38	4.08e−03	1.49	4.11e−03	1.49
346275	1.90e−03	1.33	2.29e−03	1.41	2.31e−03	1.40
825600	1.31e−03	1.27	1.56e−03	1.33	1.57e−03	1.33
1618125	9.92e−04	1.24	1.17e−03	1.28	1.18e−03	1.28
2802600	7.94e−04	1.22	9.30e−04	1.25	9.41e−04	1.25
6662400	5.62e−04	1.20	6.55e−04	1.22	6.62e−04	1.22

5.3.2.1 Regularity

The regularity results are similar to those stated in the case of a three-dimensional prismatic domain. Particularly Lemma 5.51, Corollary 5.52, Corollary 5.53 and Lemma 5.54 are also valid in this two-dimensional setting. Of course one has to substitute 3 by 2 where necessary. These results were mainly a consequence of Lemma 4.36 and some embeddings such that with Lemma 4.31 and the same embeddings the proofs work out also in the two-dimensional case.

5.3.2.2 Approximation error estimate

As in Subsection 4.4.2 we distinguish between conforming and nonconforming discretization.

Conforming elements We consider the conforming element pairs described in Subsection 4.4.2.2. Since $X_h \subset H_0^1(\Omega)^2$ for these elements the Poincaré inequality stated in Lemma 5.55 is trivially satisfied such that the boundedness of S_h can be proved as in Lemma 5.56. This means the Assumptions VAR1 and PP1 hold.

From the finite element error estimate in Lemma 4.34 one can conclude that also Assumption VAR2 holds. This means the error estimates for the variational discrete approach given in Theorem 5.7 hold in this setting. Notice, that one has to substitute $\bar{y}$ by $\bar{v}$ and $\bar{p}$ by $\bar{w}$ again.

The Assumption PP2 for the postprocessing approach follows from Lemma 4.34. The Assumptions PP3 and PP4 do actually not depend on the spaces M_h and X_h but only on the regularity of the solution and the underlying mesh. The regularity of the components of the solution of the Stokes equation is similar to that of the Poisson equation. This means that the Assumptions PP3 and PP4 can be proved similar to Lemma 5.47 and Lemma 5.48 by componentwise consideration. Consequently all Assumptions PP1–PP4 are satisfied for the postprocessing approach such that Theorem 5.14 holds with substituting $\bar{y}$ by $\bar{v}$ and $\bar{p}$ by $\bar{w}$.

Table 5.11: $L^2(\Omega)$-errors of the computed control $\tilde{u}_h$, velocity $\bar{v}_h$ and adjoint velocity $\bar{w}_h$ on graded meshes ($\mu = 0.4$) with ($\mathcal{P}_2, \mathcal{P}_0$) element.

ndof	$\|u - \tilde{u}_h\|$	eoc	$\|v - \bar{v}_h\|$	eoc	$\|w - \bar{w}_h\|$	eoc
2362	3.01e−03		3.00e−03		3.05e−03	
9722	8.04e−04	1.87	8.06e−04	1.86	8.18e−04	1.86
39442	2.09e−04	1.93	2.10e−04	1.92	2.13e−04	1.92
158882	5.32e−05	1.96	5.36e−05	1.96	5.44e−05	1.96
388877	2.20e−05	1.98	2.21e−05	1.97	2.25e−05	1.97
637762	1.35e−05	1.98	1.36e−05	1.98	1.38e−05	1.98
2555522	3.38e−06	1.99	3.41e−06	1.99	3.46e−06	1.99

Nonconforming element We can apply the same argumentation for an approximation of the velocity in the lower order Crouzeix-Raviart finite element space (4.111) and of the pressure in the space of piecewise constant functions (4.112). The only thing that has to be guaranteed is the validity of the discrete Poincaré inequality as it is stated in Lemma 5.55. But this is proved in [124, Theorem II.2.3]. The corresponding finite element error estimates for the state equation are given in Lemma 4.35. Consequently, the assertions of Theorem 5.7 and Theorem 5.14 hold also for the nonconforming discretization.

5.3.2.3 Numerical tests

We consider the same optimal control problem as in Subsection 5.3.1.3, but now in the two-dimensional domain

$$\Omega = \left\{ (r\cos\varphi, r\sin\varphi) \in \mathbb{R}^2 : 0 < r < 1, 0 < \varphi < \frac{3}{2}\pi \right\}.$$

The functions f, g and v_{d} are chosen such that

$$\bar{v} = \bar{w} = \begin{pmatrix} r^\lambda \Phi_1(\varphi) \\ r^\lambda \Phi_2(\varphi) \end{pmatrix}, \quad \bar{q} = -\bar{r} = r^{\lambda-1}\Phi_p(\varphi), \quad \bar{u} = \Pi_{[-1.0,0.1]}\left(-\frac{1}{\nu}\bar{w}\right)$$

is the exact solution of the optimal control problem. The functions Φ_1, Φ_2 and Φ_p are defined in Subsection 5.3.1.3 and $\lambda \approx 0.5445$ is again the smallest positive solution of (4.128).

In our first test we use the ($\mathcal{P}_2, \mathcal{P}_0$) element. In Table 5.11 one can find the results for appropriately graded meshes ($\mu = 0.4 < \lambda$). The predicted convergence rate of 2 can be seen in all three variables. For quasi-uniform meshes the convergence rates are significantly smaller than two, comp. Table 5.12.

The second test uses lower order Crouzeix-Raviart finite elements. From Tables 5.13 and 5.14 one realizes also second order convergence in all three variables on the graded meshes ($\mu = 0.4 < \lambda$) and a convergence rate of $2\lambda \approx 1.09$ on quasi-uniform meshes. This confirms our theoretical findings.

Table 5.12: $L^2(\Omega)$-errors of the computed control $\tilde{u}_h$, velocity $\bar{v}_h$ and adjoint velocity $\bar{w}_h$ on quasi-uniform meshes ($\mu = 1.0$) with $(\mathcal{P}_2, \mathcal{P}_0)$ element.

ndof	$\|u - \tilde{u}_h\|$	eoc	$\|v - \bar{v}_h\|$	eoc	$\|w - \bar{w}_h\|$	eoc
2362	3.25e−03		3.27e−03		3.28e−03	
9722	1.16e−03	1.46	1.17e−03	1.45	1.18e−03	1.45
39442	4.16e−04	1.46	4.24e−04	1.45	4.26e−04	1.45
158882	1.52e−04	1.44	1.57e−04	1.43	1.58e−04	1.43
388877	8.08e−05	1.41	8.41e−05	1.39	8.48e−05	1.39
637762	5.73e−05	1.39	6.00e−05	1.36	6.06e−05	1.36
2555522	2,25e−05	1.35	2.40e−05	1.32	2.43e−05	1.32

Table 5.13: $L^2(\Omega)$-errors of the computed control $\tilde{u}_h$, velocity $\bar{v}_h$ and adjoint velocity $\bar{w}_h$ on graded meshes ($\mu = 0.4$) with Crouzeix-Raviart element

ndof	$\|u - \tilde{u}_h\|$	eoc	$\|v - \bar{v}_h\|$	eoc	$\|w - \bar{w}_h\|$	eoc
1930	6.83e−03		7.36e−03		7.47e−03	
7860	1.84e−03	1.87	2.01e−03	1.85	2.04e−03	1.85
31720	4.80o−04	1.93	5.27e−04	1.92	5.34e−04	1.92
127440	1.23e−04	1.96	1.35e−04	1.96	1.37e−04	1.96
311625	5.08e−05	1.97	5.60e−05	1.97	5.67e−05	1.97
510880	3.11e−05	1.98	3.43e−05	1.98	3.48e−05	1.98
2045760	7.85e−06	1.99	8.65e−06	1.99	8.78e−06	1.99

Table 5.14: $L^2(\Omega)$-errors of the computed control $\tilde{u}_h$, velocity $\bar{v}_h$ and adjoint velocity $\bar{w}_h$ on quasi-uniform mesh ($\mu = 1.0$) with Crouzeix-Raviart element

ndof	$\|u - \tilde{u}_h\|$	eoc	$\|v - \bar{v}_h\|$	eoc	$\|w - \bar{w}_h\|$	eoc
1930	1.70e−02		1.90e−02		1.95e−02	
7860	8.23e−03	1.03	9.24e−03	1.03	9.45e−03	1.03
31720	3.91e−03	1.07	4.39e−03	1.07	4.49e−03	1.07
127440	1.84e−03	1.08	2.07e−03	1.08	2.12e−03	1.08
311625	1.13e−03	1.09	1.27e−03	1.09	1.30e−03	1.09
510880	8.63e−04	1.09	9.74e−04	1.09	9.94e−04	1.09
2045760	4.05e−04	1.09	4.57e−04	1.09	4.67e−04	1.09

CHAPTER 6

Conclusion and Outlook

In this thesis we considered PDE-constrained linear-quadratic optimal control problems with pointwise constraints on the control. Our main focus was on problems where the underlying domain had corners or edges that cause singularities in the solution. We used a priori mesh grading techniques to counteract these singularities. For the derivation of error estimates for approximations of such problems we had to deal with regularity issues, finite element error analysis and optimal control theory. To validate our results numerically we implemented the primal-dual active set strategy in C++ and coupled it with a finite element library.

We started our considerations with finite element error estimates for scalar elliptic equations and the Stokes equation. Here we first focussed on estimates of the pointwise error for scalar elliptic equations in domains with corners. We proved a convergence rate of $h^2 |\ln h|^{3/2}$ on graded meshes for problems with Hölder continuous right-hand side. It turned out that mesh grading is necessary also in convex domains with an interior angle larger than $\pi/2$. The exact classification of the regularity of the right-hand side was new and made the result applicable to optimal control problems. A prerequisite was the proof of a regularity result in a weighted Sobolev space for such data. In a next step we considered elliptic equations with pure Dirichlet and Neumann boundary conditions in prismatic domains with reentrant edge. For the derivation of finite element error estimates on anisotropic meshes we had to use quasi-interpolation operators. As such an operator has to preserve the boundary conditions for the Dirichlet problem we introduced a modification of the operator E_h defined in [6] and derived the corresponding local and global estimates. For the Neumann problem we could use the operator E_h. But the solution of the Neumann problem admits different regularity properties than the examples treated in [6] and we had to adapt the proofs for that case. For the Stokes equations we went a slightly different way. We first stated a couple of general assumptions that allowed to prove optimal error estimates. Afterwards we verified these assumptions for problems in nonconvex domains with reentrant corner or edge. In a two-dimensional setting we used isotropic graded meshes and proved the assumptions for discretizations with several well-known conforming element pairs as well as for a discretization of the velocity space with Crouzeix-Raviart elements. The same non-conforming approximation was used in a prismatic domain with reentrant edge and anisotropic graded mesh.

We did the error analysis for the optimal control problems for a general linear-quadratic case first. We considered the variational discrete approach introduced by Hinze [71] and the post-processing approach introduced by Meyer and Rösch [95]. For both approaches we gave a couple of assumptions that were sufficient to prove second order convergence in $L^2(\Omega)$ in all three variables, i.e., control, state and adjoint state. We checked these assumptions for scalar elliptic equations with pure Dirichlet or Neumann boundary conditions in a prismatic domain with reentrant edge and showed that the estimates hold for piecewise linear approximations of state and adjoint state on anisotropic graded meshes with the same grading condition as necessary for optimal convergence in the state equation itself. We got this result on the same graded meshes also for the Stokes equations as state equation and an approximation in the Crouzeix-Raviart finite element space. The Stokes equations as state equation were also treated for a optimal control problem in a two-dimensional domain with reentrant corner. We checked the assumptions for a couple of element pairs on an isotropic graded mesh. In such a domain we also considered an example with a state equation that has nonsmooth coefficients. We showed that such a configuration also fits in the general framework and proved the corresponding error estimates. The case of pointwise error estimates for problems with scalar elliptic state equation were treated separately. We could show that the convergence rates for the optimal control problem on appropriately graded meshes are the same as for the boundary value problem. All the results concerning the post-processing approach were confirmed by numerical tests.

Let us briefly discuss some possible extensions of our results. We have treated linear state equations only. Here, one can think of a generalization to different types of nonlinear equations. Of special interest might be an optimal control problem with the incompressible Navier-Stokes equations as state equation in domains with edges and its discretization with a nonconforming finite element method on anisotropic graded meshes. The boundary value problem is investigated in [82], the extension to optimal control is an open question. We restricted our considerations to the control constrained case. A natural extension would be the treatment of pointwise constraints on the state. In recent years many publications were devoted to the numerical analysis of such problems, see, e.g., [41, 51, 52, 53, 72, 93], but all of them are restricted to convex domains. Although the analysis differs significantly from the control constrained case a very important ingredient are L^∞-error estimates. Since the control is in $L^\infty(\Omega)$ but not necessarily in $C^{0,\sigma}(\Omega)$ the estimates of Section 4.1 are not applicable. A generalization of these results to right-hand sides in $L^\infty(\Omega)$ is not straightforward. As a third possible extension let us mention boundary control problems. We considered only examples with distributed control. An interesting question is how one can use mesh grading techniques for boundary control problems, especially in three dimensions and on anisotropic meshes.

APPENDIX A

Notation

symbols	description	pages
$\sim$	$a \sim b \Leftrightarrow \exists c_1, c_2 \in \mathbb{R} : c_1 a \le b \le c_2 a$	
$\alpha, |\alpha|$	multiindex $\alpha = (\alpha_1, \alpha_2, \alpha_3) \in \mathbb{N}_0^3$ with $|\alpha| := \sum_{i=1}^{3} \alpha_i$	
c	generic constant independent of the mesh size h; it may have a different value on each occurrence	
$C^{0,\sigma}(\Omega)$	space of Hölder continuous functions	12
∂_i	first order derivative w.r.t. the i-th variable	
∂_{ij}	second order derivative w.r.t. the i-th and j-th variable	
D^α	differential operator $D^\alpha := \frac{\partial^{\alpha_1}}{\partial x_1^{\alpha_1}} \frac{\partial^{\alpha_2}}{\partial x_1^{\alpha_2}} \frac{\partial^{\alpha_3}}{\partial x_1^{\alpha_3}}$	
eoc	estimated order of convergence	
E_h, E_{0h}	modified Scott-Zhang interpolation operators	22, 23
h^α	multiindex notation; $h^\alpha := h_1^{\alpha_1} h_2^{\alpha_2} h_3^{\alpha_3}$	
$H^k(\Omega)$	Sobolev space	11
I_h	Lagrangian (or nodal) interpolation operator	
λ	singularity exponent	36, 64, 71, 79
M_T	certain patch of elements	21
$\Omega_j, \Omega_j', \Omega_j''$	subsets of Ω	39
$\Pi_{U^{\text{ad}}}$	projection in the space of admissible controls	84
$\mathcal{P}_k$	space of continuous functions that are (piecewise) polynomials of order at most k	

symbols	description	pages
P, P_h	(affine) operator that maps a given control to the corresponding adjoint state and its discrete version	
S	solution operator of the state equation	
S^*	adjoint operator of S	
S_h, S_h^*	discretizations of S and S^*	
S_T	certain patch of elements	21
$\mathcal{T}_h$	admissible triangulation of Ω	
$T, \bar{T}, |T|$	finite element, its closure and its measure	
$\hat{T}$	reference element	
U^{ad}	set of admissible controls	
$V_\beta^{k,p}(\Omega)$	weighted Sobolev space	11
$W_\beta^{k,p}(\Omega)$	weighted Sobolev space	11
$W^{k,p}(\Omega)$	classical Sobolev space	11

The Software Package OPTPDE

All the numerical tests in this thesis were computed with the software package OPTPDE implemented in C++ by the author. In this chapter we describe the structure of the software and explain the implemented algorithms.

B.1 Structure of OPTPDE

The principle structure of OPTPDE is illustrated in Figure B.1. The package consists of three parts, the finite element library, the interfaces and the optimization part. The innermost part is the finite element library. The task of this library is to manage the meshes and to assemble the stiffness and mass matrices. The interface provides particular methods which makes a communication between optimization and finite element part possible. The algorithm for computing the solution of the optimal control problem is situated in the optimization part. The advantage of such a structure is the independency of finite element and optimization part of the code. The optimization part communicates with the finite element library through interfaces such that one could use different finite element libraries and would just have to adapt the interface. The optimization code could keep untouched.

B.2 Optimization part

The main ingredient of the optimization part of OPTPDE is an implementation of the primal-dual active set strategy to solve control-constrained optimal control problems. A detailed analysis of this algorithm can be found in [81]. In the following we describe how its implemented in our software package. We want to solve the optimality system which is

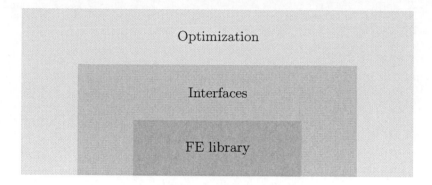

Figure B.1: Structure of software package `OPTPDE`

given in Subsection 5.1.2.2,

$$\bar{y}_h = S_h \bar{u}_h,$$
$$\bar{p}_h = S_h^*(\bar{y}_h - y_d),$$
$$(\nu \bar{u}_h + \bar{p}_h, u_h - \bar{u}_h)_U \geq 0 \quad \forall u_h \in U_h^{\text{ad}}.$$

Let $T_h = \{T_1, \ldots, T_n\}$ be a triangulation of Ω. For a piecewise constant function $u_h \in U_h$ we introduce three sets of elements, namely the active sets $\mathcal{A}^+$, $\mathcal{A}^-$ and the inactive set $\mathcal{I}$,

$$\mathcal{A}^- = \{T \in T_h : u_h(x) = u_a \ \forall x \in T\},$$
$$\mathcal{A}^+ = \{T \in T_h : u_h(x) = u_b \ \forall x \in T\}, \tag{B.1}$$
$$\mathcal{I} = \{T \in T_h : u_a < u_h(x) < u_b \ \forall x \in T\}.$$

Since u_h is piecewise constant one has $\Omega = \mathcal{A}^- \cup \mathcal{A}^+ \cup \mathcal{I}$. The algorithm constructs a sequence of sets of active and inactive elements. This sequence approximates the active and inactive set of the continuous problem. The principal structure of the primal-dual active set strategy is given in Algorithm B.1.

In line 2 of Algorithm B.1 one has to solve the optimality system of the unconstrained optimal control problem

$$\min_{u_h^k \in U_h} \frac{1}{2} \|S_h u_h^k - y_d\|_{L^2(\Omega)}^2 + \frac{\nu}{2} \|u_h^k\|_{L^2(\Omega)}$$
$$u_h^k = u_a \text{ on } \mathcal{A}_k^-, \quad u_h^k = u_b \text{ on } \mathcal{A}_k^+.$$

This represents the idea of the primal-dual active set strategy, namely the approximation of a constrained optimal control problem by a sequence of unconstrained problems. For the solution of the system in line 2 of Algorithm B.1 we introduce

$$u_{\mathcal{A}}^k = \begin{cases} u_a & \text{if } x \in \mathcal{A}_k^- \\ 0 & \text{if } x \in \mathcal{I} \\ u_b & \text{if } x \in \mathcal{A}_k^+ \end{cases} \quad \text{and} \quad u_{\mathcal{I}}^k = u_h^k - u_{\mathcal{A}}^k.$$

Algorithm B.1 Primal-dual active set strategy as implemented in OPTPDE

1: Initialization: $k := 0$, $\mathcal{A}_0^- = \emptyset$, $\mathcal{A}_0^+ = \emptyset$
2: Compute u_h^k, y_h^k and p_h^k from the system

$$y_h^k = S_h u_h^k,$$
$$p_h^k = S_h^*(y_h^k - y_d),$$
$$u_h^k = \begin{cases} u_a & \text{if } x \in \mathcal{A}_k^-, \\ u_b & \text{if } x \in \mathcal{A}_k^+, \\ -\frac{1}{\nu} p_h^k & \text{if } x \in \mathcal{I}_k. \end{cases}$$

3: Determine the new sets $\mathcal{A}_{k+1}^-$, $\mathcal{A}_{k+1}^+$ and $\mathcal{I}_{k+1}$ according to (B.1).
4: If $\mathcal{A}_k^- \neq \mathcal{A}_{k+1}^-$ or $\mathcal{A}_k^+ \neq \mathcal{A}_{k+1}^+$ then set $k := k + 1$ and go to step 2.

This allows to reformulate line 2 as

$$(\nu \text{Id} + S_h^* S_h) u_{\mathcal{I}}^k = -S_h^*(S_h u_{\mathcal{A}}^k - y_d) - \nu u_{\mathcal{A}}^k.$$

For the solution of this equation we use a simple preconditioned cg-algorithm as its described in Algorithm B.2. In our case one has

$$B = \nu \text{Id} + S_h^* S_h \quad \text{and} \quad b = -S_h^*(S_h u_{\mathcal{A}}^k - y_d) - \nu u_{\mathcal{A}}^k,$$

which means that in line 8 of Algorithm B.2 one has to solve two boundary value problems.

Algorithm B.2 A simple preconditioned cg-algorithm for the solution of $Bu = b$

1: Choose a start value u and a invertible preconditioner matrix M
2: $r := b - Bu$
3: $g := M^{-1} r$
4: $\varepsilon = \sqrt{(r,g)}$
5: **while** $\sqrt{(r, M^{-1}r)}/\varepsilon > TOL$ **do**
6: $\quad z := M^{-1} r$
7: $\quad \alpha := (r, z)$
8: $\quad d := Bg$
9: $\quad \gamma := (d, g)$
10: $\quad u := u + \frac{\alpha}{\gamma} u$
11: $\quad r := r - \frac{\alpha}{\gamma} d$
12: $\quad \beta := (r, z)$
13: $\quad g := z + \frac{\beta}{\alpha} g$
14: **end while**

To deal with matrices and vectors efficiently we have integrated the sparse matrix class library SparseLib++ [107]. For the solution of the occurring linear systems we offer direct as well as iterative solvers. For the iterative solvers we use the iterative methods library IML++ [56] which is tailored to SparseLib++. In our program one can choose between a BiCGSTAB and a GMRES solver. As direct solver we have included UMFPACK [48] and PARDISO [119]. During the solution process one has to solve the same linear system with different right-hand sides several times. Here, one can profit from a direct solver since one can reuse the factorization and therefore reduce computational costs significantly. For this reason we used PARDISO in all our examples which is much less memory consuming than UMFPACK. The discretization of a Stokes problem results in a saddle point problem of type

$$\begin{pmatrix} A & C^T \\ C & 0 \end{pmatrix} \begin{pmatrix} v \\ q \end{pmatrix} = \begin{pmatrix} f \\ 0 \end{pmatrix} \tag{B.2}$$

with matrices A and C and vectors v and q. To solve this system we have implemented a simple preconditioned Uzawa algorithm. This means, the system

$$CA^{-1}C^T q = CA^{-1}f,$$

which results from expressing v with help of the first row of (B.2) and then plugging it in the second row, is solved for q by Algorithm B.2 ($B := CA^{-1}C^T$, $b := CA^{-1}f$). By adding the expression $v := v - \alpha/\gamma\, A^{-1}C^T g$ after line 11 in that algorithm one can get v nearly without any additional effort since $A^{-1}C^T g$ is already computed in line 8. The linear systems with matrix A can be solved with all the solvers mentioned above. For the numerical tests in Chapter 5 we used PARDISO and profited from the reuseable factorization of A during the iteration.

B.3 Finite element library

We use MoonMD [75] as finite element library. This library offers many different element types for a couple of equations. It returns mass and stiffness matrices in a format that is compatible to the one used in the matrix library SparseLib++. For the computation of the error norms $\|\bar{y} - \bar{y}_h\|_{L^2(\Omega)}$ and $\|\bar{p} - \bar{p}_h\|_{L^2(\Omega)}$ we can directly use the routines implemented in MoonMD with appropriate integration rules. This is not the case for the evaluation of the error $\|\bar{u} - \tilde{u}_h\|_{L^2(\Omega)}$. The reason is that the optimal control $\bar{u}$ and its approximation $\tilde{u}_h$ admit kinks due the projection formulas (5.6) and (5.14) and that these kinks in general do not fit the mesh. Thus triangles/tetrahedra T where $\bar{u}$ is not smooth or $\tilde{u}_h|_T$ is not linear request some special treatment. Therefore we implemented Algorithm B.3 for the computation of $\|\bar{u} - \tilde{u}_h\|_{L^2(\Omega)}$.

Algorithm B.3 Computation of $\|\bar{u} - \tilde{u}_h\|_{L^2(\Omega)}$

1: $\|\bar{u} - \tilde{u}_h\|_{L^2(\Omega)} = 0$
2: **for** $i = 0\ldots$ number of elements **do**
3:     compute $e_1 := \|\bar{u} - \tilde{u}_h\|_{L^2(T_i)}$ using a standard integration rule
4:     **if** $\bar{u}$ or $\tilde{u}_h$ is not smooth on T_i **then**
5:         **for** $k = 0\ldots$ MAXLEVEL **do**
6:             refine T_i or all its subelements into subelements τ_j
7:             compute $e_2 := \sum_j \|\bar{u} - \tilde{u}_h\|_{L^2(\tau_j)}^2$ via standard integration rule on every τ_j
8:             **if** $|e_1 - \sqrt{e_2}| < $ RELTOL $\cdot e_1$ **then**
9:                 $e_1 := \sqrt{e_2}$
10:                 **break**
11:             **else**
12:                 $e_1 := \sqrt{e_2}$
13:             **end if**
14:         **end for**
15:     **end if**
16:     $\|\bar{u} - \tilde{u}_h\|_{L^2(\Omega)} := \sqrt{\|\bar{u} - \tilde{u}_h\|_{L^2(\Omega)}^2 + e_1^2}$
17: **end for**

Bibliography

[1] H. W. Alt. *Lineare Funktionalanalysis*. Springer, 2006.

[2] I. Altrogge, T. Preusser, T. Kröger, C. Büskens, P. L. Pereira, D. Schmidt, and H.-O. Peitgen. Multiscale optimization of the probe-placement for radio-frequency ablation. *Acad. Radiol.*, 14(11):1310–1324, 2007.

[3] T. Apel, O. Benedix, D. Sirch, and B. Vexler. A priori mesh grading for an elliptic problem with Dirac right-hand side. Preprint SPP1253-087, DFG Priority Program 1253, Erlangen, 2009. submitted.

[4] T. Apel and T. G. Flaig. Simulation and mathematical optimization of the hydration of concrete for avoiding thermal cracks. In K. Gürlebeck and C. Könke, editors, *Proceedings of the 18th International Conference on the Applications of Computer Science and Mathematics in Architecture and Civil Engineering*, Weimar, 2009.

[5] T. Apel, J. Pfefferer, A. Rösch, and D. Sirch. Corrigendum: L^∞-error estimates on graded meshes with application to optimal control. *in preparation*, 2010.

[6] Th. Apel. Interpolation of non-smooth functions on anisotropic finite element meshes. *Math. Modeling Numer. Anal.*, 33:1149–1185, 1999.

[7] Th. Apel and M. Dobrowolski. Anisotropic interpolation with applications to the finite element method. *Computing*, 47:277–293, 1992.

[8] Th. Apel and B. Heinrich. Mesh refinement and windowing near edges for some elliptic problem. *SIAM J. Numer. Anal.*, 31:695–708, 1994.

[9] Th. Apel and S. Nicaise. Elliptic problems in domains with edges: anisotropic regularity and anisotropic finite element meshes. Preprint SPC94_16, TU Chemnitz-Zwickau, 1994.

[10] Th. Apel and S. Nicaise. Elliptic problems in domains with edges: anisotropic regularity and anisotropic finite element meshes. In J. Cea, D. Chenais, G. Geymonat, and J. L. Lions, editors, *Partial Differential Equations and Functional Analysis (In Memory of Pierre Grisvard)*, pages 18–34. Birkhäuser, Boston, 1996. Shortened version of Preprint SPC94_16, TU Chemnitz-Zwickau, 1994.

[11] Th. Apel and S. Nicaise. The finite element method with anisotropic mesh grading

for elliptic problems in domains with corners and edges. *Math. Methods Appl. Sci.*, 21:519–549, 1998.

[12] Th. Apel, S. Nicaise, and J. Schöberl. Crouzeix-Raviart type finite elements on anisotropic meshes. *Numer. Math.*, 89:193–223, 2001.

[13] Th. Apel, S. Nicaise, and J. Schöberl. A non-conforming finite element method with anisotropic mesh grading for the Stokes problem in domains with edges. *IMA J. Numer. Anal.*, 21:843–856, 2001.

[14] Th. Apel, A. Rösch, and D. Sirch. L^∞-error estimates on graded meshes with application to optimal control. *SIAM J. Control Optim.*, 48(3):1771–1796, 2009.

[15] Th. Apel, A. Rösch, and G. Winkler. Optimal control in non-convex domains: a priori discretization error estimates. *Calcolo*, 44:137–158, 2007.

[16] Th. Apel, A.-M. Sändig, and J. R. Whiteman. Graded mesh refinement and error estimates for finite element solutions of elliptic boundary value problems in non-smooth domains. *Math. Methods Appl. Sci.*, 19:63–85, 1996.

[17] Th. Apel and D. Sirch. L^2-error estimates for the Dirichlet and Neumann problem on anisotropic finite element meshes. Preprint SPP1253-02-05, DFG Priority Program 1253, Erlangen, 2008. Accepted for publication in Appl. Math.

[18] Th. Apel, D. Sirch, and G. Winkler. Error estimates for control contstrained optimal control problems: Discretization with anisotropic finite element meshes. Preprint SPP1253-02-06, DFG Priority Program 1253, Erlangen, 2008.

[19] Th. Apel and G. Winkler. Optimal control under reduced regularity. *Appl. Numer. Math.*, 59:2050–2064, 2009.

[20] N. Arada, E. Casas, and F. Tröltzsch. Error estimates for the numerical approximation of a semilinear elliptic control problem. *Comput. Optim. Appl.*, 23:201–229, 2002.

[21] I. Babuška. The finite element method for elliptic equations with discontinuous coefficients. *Computing*, 5:207–213, 1970.

[22] I. Babuška, R. B. Kellogg, and J. Pitkäranta. Direct and inverse error estimates for finite elements with mesh refinements. *Numer. Math.*, 33:447–471, 1979.

[23] I. Babuška, T. von Petersdorff, and B. Andersson. Numerical treatment of vertex singularities and intensity factors for mixed boundary value problems for the Laplace equation in $\mathbb{R}^3$. *SIAM J. Numer. Anal.*, 31:1265–1288, 1994.

[24] G. Baker. *Projection methods for boundary value problems for equations of elliptic and parabolic type with discontinuous coefficients.* PhD thesis, Cornell University, 1973.

[25] J. Banasiak and G. F. Roach. On mixed boundary value problems of Dirichlet oblique-derivative type in plane domains with piecewise differentiable boundary. *J. Differential Equations*, 79:111–131, 1989.

[26] J. W. Barrett and C. M. Elliott. Fitted and unfitted finite element methods for elliptic equations with smooth interfaces. *IMA J. Numer. Math.*, 7:283–300, 1987.

[27] A. E. Beagles and J. R. Whiteman. Finite element treatment of boundary singularities by augmentation with non-exact singular functions. *Numer. Methods Partial Differential Equations*, 2:113–121, 1986.

[28] O. Benedix and B. Vexler. A posteriori error estimation and adaptivity for elliptic optimal control problems with state constraints. *Comput. Optim. Appl.*, 44(1):3–25, 2009.

[29] H. Blum and M. Dobrowolski. On finite element methods for elliptic equations on domains with corners. *Computing*, 28:53–63, 1982.

[30] P. Bochev and M. Gunzburger. Least-squeares finite-element methods for optimization and control problemes for the Stokes equations. *Comput. Math. Appl.*, 48:1035–1057, 2004.

[31] M. Bourlard, M. Dauge, J. M.-S. Lubuma, and S. Nicaise. Coefficients of the singularities for elliptic boundary value problems on domain with conical points III: Finite elements methods on polygonal domains. *SIAM J. Numer. Anal.*, 29:136–155, 1992.

[32] D. Braess. *Finite Elemente*. Springer, Berlin, 1997.

[33] J. H. Bramble and J. T. King. A finite element method for interface problems in domains with smooth boundaries and interfaces. *Adv. Comput. Math.*, 6:109–138, 1996.

[34] S. C. Brenner and L. R. Scott. *The mathematical theory of finite element methods*. Springer, New York, 1994.

[35] F. Brezzi and M. Fortin. *Mixed and hybrid finite element methods*. Springer, New York, 1991.

[36] E. Casas. Using piecewise linear functions in the numerical approximation of semilinear elliptic control problems. *Adv. Comput. Math.*, 26:137–153, 2007.

[37] E. Casas, M. Mateos, and J.-P. Raymond. Error estimates for the numerical approximation of a distributed control problem for the steady-state Navier-Stokes equations. *SIAM J. Control Optim.*, 46:952–982, 2007.

[38] E. Casas, M. Mateos, and F. Tröltzsch. Error estimates for the numerical approximation of boundary semilinear elliptic control problems. *Comput. Optim. Appl.*, 31:193–219, 2005.

[39] E. Casas and F. Tröltzsch. Error estimates for linear-quadratic elliptic control problems. In V. Barbu et al., editor, *Analysis and optimization of differential systems.*, pages 89–100, Boston, MA, 2003. Kluwer Academic Publisher.

[40] Y. Chen. Superconvergence of mixed finite element methods for optimal control problems. *Math. Comp.*, 77(263):1269–1291, 2008.

[41] S. Cherednichenko and A. Rösch. Error estimates for the discretization of elliptic control problems with pointwise control and state constraints. *Comput. Optim. Appl.*, 44(1):27–55, 2009.

[42] P. G. Ciarlet. *The finite element method for elliptic problems.* North-Holland, Amsterdam, 1978. Reprinted by SIAM, Philadelphia, 2002.

[43] P. G. Ciarlet and J.-P. Lions, editors. *Handbook of Numerical Analysis, Volume II: Finite Element Method (Part 1)*, Amsterdam, 1991. North Holland.

[44] P. Clément. Approximation by finite element functions using local regularization. *RAIRO Anal. Numer.*, 2:77–84, 1975.

[45] M. Dauge. *Elliptic boundary value problems on corner domains – smoothness and asymptotics of solutions*, volume 1341 of *Lecture Notes in Mathematics*. Springer, Berlin, 1988.

[46] M. Dauge. Stationary Stokes and Navier-Stokes systems on two- and three-dimensional domains with corners. Part I: Linearized equations. *SIAM J. Math. Anal.*, 20:27–52, 1989.

[47] M. Dauge. Neumann and mixed problems on curvilinear polyhedra. *Integr. Equat. Oper. Th.*, 15:227–261, 1992.

[48] T. Davis. Algorithm 832: Umfpack v4.3–an unsymmetric-pattern multifrontal method. *ACM Trans. Math. Software*, 30(2):196–199, 2004.

[49] K. Deckelnick, A. Günther, and M. Hinze. Finite element approximation of elliptic control problems with constraints on the gradient. *Numer. Math.*, 111(3):335–350, 2009.

[50] K. Deckelnick and M. Hinze. Semidiscretization and error estimates for distributed control of the instationary Navier-Stokes equations. *Numer. Math.*, 97:297–320, 2004.

[51] K. Deckelnick and M. Hinze. Convergence of a finite element approximation to state constrained elliptic control problem. *SIAM J. Numer. Anal.*, 45(5):1937–1953, 2007.

[52] K. Deckelnick and M. Hinze. A finite element approximation to elliptic control problems in the presence of control and state constraints. *Preprint HBAM2007-01, Hamburger Beiträge zur Angewandten Mathematik*, 2007.

[53] K. Deckelnick and M. Hinze. Numercal analysis of a control and state contstrained elliptic control problem with piecewise constant control approximations. *Preprint HBAM2007-02, Hamburger Beiträge zur Angewandten Mathematik*, 2007.

[54] M. Dobrowolski. Numerical approximation of elliptic interface and corner problems. Habilitationsschrift, Universität Bonn, 1981.

[55] G. Dolzmann and S. Müller. Estimates for Green's matrices of elliptic systems by L^p theory. *Manuscripta Math.*, 88:261–273, 1995.

[56] J. Dongarra, A. Lumsdaine, R. Pozo, and K. A. Remington. Iml++ v. 1.2 reference

guide. http://math.nist.gov/iml++/.

[57] M. Falk. Approximation of a class of optimal control problems with order of convergence estimates. *J. Math. Anal. Appl.*, 44:28–47, 1973.

[58] J. Frehse and R. Rannacher. Eine L^1-Fehlerabschätzung für diskrete Grundlösungen in der Methode der finiten Elemente. In J. Frehse, R. Leis, and R. Schaback, editors, *Finite Elemente. Tagungsband des Sonderforschungsbereichs 72*, volume 89 of *Bonner Mathematische Schriften*, pages 92–114, Bonn, 1976.

[59] A. Gaevskaya, R. H. W. Hoppe, Y. Iliash, and M. Kieweg. Convergence analysis of an adaptive finite element method for distributed control problems with control constraints. In *Control of Coupled Partial Differential Equations*, pages 22–48, Basel, 2007. Birkhäuser.

[60] T. Geveci. On the approximation of the solution of an optimal control problem governed by an elliptic equation. *RAIRO, Anal. Numér.*, 13:313–328, 1979.

[61] V. Girault and P.-A. Raviart. *Finite element methods for Navier-Stokes equations. Theory and algorithms*, volume 5 of *Springer Series in Computational Mathematics*. Springer, Berlin, 1986.

[62] P. Grisvard. *Elliptic problems in nonsmooth domains*, volume 24 of *Monographs and Studies in Mathematics*. Pitman, Boston, 1985.

[63] P. Grisvard. *Singularities in boundary value problems*, volume 22 of *Research Notes in Applied Mathematics*. Springer, New York, 1992.

[64] P. Grisvard. Singular behaviour of elliptic problems in non Hilbertian Sobolev spaces. *J. Math. Pures Appl.*, 74:3 33, 1995.

[65] A. Günther and M. Hinze. Elliptic control problems with gradient constraints - variational discrete versus piecewise constant controls. Preprint SPP1253-08-07, DFG Priority Program 1253, Erlangen, 2008. accepted for publication in Comput. Optim. Appl.

[66] A. Günther and M. Hinze. A posteriori error control of a state constrained elliptic control problem. *J. Numer. Math.*, 16(4):307–322, 2008.

[67] M. Gunzburger, L. Hou, and T. Svobodny. Analysis and finite element approximation of optimal control problems for the stationary Navier-Stokes equations with Dirichlet controls. *Math. Model. Numer. Anal.*, 25:711–748, 1991.

[68] M. Gunzburger, L. Hou, and T. Svobodny. Analysis and finite element approximation of optimal control problems for the stationary Navier-Stokes equations with distributed and Neumann controls. *Math. Comp.*, 57:123–151, 1991.

[69] M. Hintermüller and R. H. W. Hoppe. Goal-oriented adaptivity in control constrained optimal control of partial differential equations. *SIAM J. Control Optim.*, 47:1721–1743, 2008.

[70] M. Hintermüller, R. H. W. Hoppe, Y. Iliash, and M. Kieweg. An a posteriori error

analysis of adaptive finite element methods for distributed elliptic control problems with control constraints. *ESAIM: Control Optim. Calc. Var.*, 14:540–560, 2008.

[71] M. Hinze. A variational discretization concept in control constrained optimization: The linear-quadratic case. *Comput. Optim. Appl.*, 30:45–61, 2005.

[72] M. Hinze and C. Meyer. Variational discretization of Lavrientiev-regularized state constrained elliptic optimal control problems. *Comput. Optim. Appl.*, DOI 10.1007/s10589-008-9198-1, 2008. electronically published.

[73] M. Hinze, R. Pinnau, M. Ulbrich, and S. Ulbrich. *Optimization with PDE Constraints*. Springer, 2008.

[74] R. H. W. Hoppe and M. Kieweg. Adaptive finite element methods for mixed control-state constrained optimal control problems for elliptic boundary value problems. *Comput. Optim. Appl.*, DOI 0.1007/s10589-008-9195-4, 2008. electronically published.

[75] V. John and G. Matthies. MooNMD – a program package based on mapped finite element methods. *Comput. Vis. Sci.*, 6:163–169, 2004.

[76] J. T. King. A quasioptimal finite element method for elliptic interface problems. *Computing*, 15:127–135, 1975.

[77] V. A. Kondrat'ev. Boundary value problems for elliptic equations on domains with conical or angular points. *Trudy Moskov. Mat. Obshch.*, 16:209–292, 1967. In Russian.

[78] V. A. Kozlov, V. G. Maz'ya, and J. Roßmann. *Elliptic Boundary Value Problems in Domains with Point Singularities*. American Mathematical Society, Providence, RI, 1997.

[79] V. A. Kozlov, V. G. Maz'ya, and J. Roßmann. *Spectral Problems Associated with Corner Singularities of Solutions to Elliptic Equations*. American Mathematical Society, Providence, RI, 2001.

[80] A. Kufner and A.-M. Sändig. *Some Applications of Weighted Sobolev Spaces*. Teubner, Leipzig, 1987.

[81] K. Kunisch and A. Rösch. Primal-dual active set strategy for a general class of constrained optimal control problems. *SIAM J. Optim.*, 13:321–334, 2002.

[82] J. Lazaar and S. Nicaise. A non-conforming finite element method with anisotropic mesh grading for the incompressible Navier-Stokes equations in domains with edges. *Calcolo*, 39:123–168, 2002.

[83] R. Li, W. Liu, H. Ma, and T. Tang. Adaptive finite element approximation for distributed elliptic optimal control problems. *SIAM J. Control Optim.*, 13:321–334, 2002.

[84] J. L. Lions. *Optimal Control of Systems Governed by Partial Differential Equations*. Springer, 1971. Translation of the French edition "Contrôle optimal de systèmes

gouvernés par de équations aux dérivées partielles", Dunod and Gauthier-Villars, 1968.

[85] W. Liu and N. Yan. A posteriori error estimates for distributed convex optimal control problems. *Adv. Comput. Math.*, 15:285–309, 2001.

[86] J. M.-S. Lubuma and S. Nicaise. Dirichlet problems in polyhedral domains II: approximation by FEM and BEM. *J. Comp. Appl. Math.*, 61:13–27, 1995.

[87] J. M.-S. Lubuma and S. Nicaise. Finite element method for elliptic problems with edge singularities. *J. Comp. Appl. Math.*, 106:145–168, 1999.

[88] K. Malanowski. Convergence of approximations vs. regularity of solutions for convex, control-constrained optimal-control problems. *Appl. Math. Optim.*, 8:69–95, 1981.

[89] V. G. Maz'ya and B. A. Plamenevsky. The first boundary value problem for classical equations of mathematical physics in domains with piecewise smooth boundaries, part I, II. *Z. Anal. Anwend.*, 2:335–359, 523–551, 1983. In Russian.

[90] V. G. Maz'ya and B. A. Plamenevsky. Weighted spaces with nonhomogeneous norms and boundary value problems in domains with conical points. *Trans. Amer. Math. Soc.*, 123:89–107, 1984.

[91] V. G. Maz'ya and J. Rossmann. Schauder estimates for solutions to boundary value problems for second order elliptic systems in polyhedral domains. *Applicable Analysis*, 83:271–308, 2004.

[92] V. G. Maz'ya and J. Rossmann. Schauder estimates for solutions to a mixed boundary value problem for the Stokes system in polyhedral domains. *Math. Meth. Appl. Sci.*, 29:965–1017, 2006.

[93] C. Meyer. Error estimates for the finite element approximation of an elliptic control problem with pointwise state and control constraints. *Control Cybernet.*, 37:51–85, 2008.

[94] C. Meyer, J.C. de los Reyes, and B. Vexler. Finite element error analysis for state-constrained optimal control of the Stokes equations. *Control Cybernet.*, 37:251–284, 2008.

[95] C. Meyer and A. Rösch. Superconvergence properties of optimal control problems. *SIAM J. Control Optim.*, 43:970–985, 2004.

[96] C. Meyer and A. Rösch. L^∞-estimates for approximated optimal control problems. *SIAM J. Control and Optimization*, 44:1636–1649, 2005.

[97] S. G. Michlin. *Partielle Differentialgleichungen in der mathematischen Physik.* Akademie-Verlag, Berlin, 1978.

[98] F. Natterer. Über die punktweise Konvergenz Finiter Elemente. *Numer. Math.*, 25:67–77, 1975.

[99] S. Nazarov and B. A. Plamenevsky. *Elliptic Problems in Domains with Piecewise Smooth Boundaries.* Walther de Gruyter & Co., 1994.

[100] S. Nicaise. *Polygonal Interface Problems*, volume 39 of *Methoden und Verfahren der mathematischen Physik*. Peter Lang GmbH, Europäischer Verlag der Wissenschaften, Frankfurt/M., 1993.

[101] S. Nicaise. Regularity of the solutions of elliptic systems in polyhedral domains. *Bulletin Belgium Math. Soc.-S. Stevin*, 4:411–429, 1997.

[102] S. Nicaise and D. Sirch. Optimal control of the stokes equations: Conforming and non-conforming finite element methods under reduced regularity. *Comput. Optim. Appl.*, DOI 10.1007/s10589-009-9305-y, 2009. electronically published.

[103] L. A. Oganesyan and L. A. Rukhovets. Variational-difference schemes for linear second-order elliptic equations in a two-dimensional region with piecewise smooth boundary. *Zh. Vychisl. Mat. Mat. Fiz.*, 8:97–114, 1968. In Russian. English translation in USSR Comput. Math. and Math. Phys., 8 (1968) 129–152.

[104] L. A. Oganesyan, L. A. Rukhovets, and V. Ja. Rivkind. *Variational-difference methods for solving elliptic equations, Part II*, volume 8 of *Differential equations and their applications*. Izd. Akad. Nauk Lit. SSR, Vilnius, 1974. In Russian.

[105] C. Ortner and W. Wollner. A priori error estimates for optimal control problems with pointwise constraints on the gradient of the state. Preprint SPP1253-23-03, DFG Priority Program 1253, Erlangen, 2009.

[106] M. Petzoldt. *Regularity and error estimates for elliptic problems with discontinuous coefficients*. PhD thesis, FU Berlin, 2001.

[107] R. Pozo, K. Remington, and A. Lumsdaine. Sparselib++ v. 1.5 reference guide. http://www.math.nist.gov/sparselib++/.

[108] R. Rannacher. Numerische Mathematik 2 (Numerik Partieller Differentialgleichungen). Lecture Notes, Institut für angewandte Mathematik, Universität Heidelberg, 2008. http://numerik.iwr.uni-heidelberg.de/~lehre/notes/.

[109] G. Raugel. *Résolution numérique de problèmes elliptiques dans des domaines avec coins*. PhD thesis, Université de Rennes, 1978.

[110] G. Raugel. Résolution numérique par une méthode d'éléments finis du problème de Dirichlet pour le Laplacien dans un polygone. *C. R. Acad. Sci. Paris, Sér. A*, 286(18):A791–A794, 1978.

[111] A. Rösch. Error estimates for linear-quadratic control problems with control constraints. *Optim. Methods Softw.*, 21(1):121–134, 2006.

[112] A. Rösch and B. Vexler. Optimal control of the stokes equations: A priori error analysis for finite element discretization with postprocessing. *SIAM J. Numer. Anal.*, 44(5):1903–1920, 2006.

[113] J. Roßmann. Gewichtete Sobolev–Slobodetskiĭ–Räume und Anwendungen auf elliptische Randwertaufgaben in Gebieten mit Kanten. Habilitationsschrift, Universität Rostock, 1988.

[114] J. Roßmann. The asymptotics of the solutions of linear elliptic variational problems in domains with edges. *Z. Anal. Anwend.*, 9(3):565–575, 1990.

[115] A.-M. Sändig. Error estimates for finite element solutions of elliptic boundary value problems in non-smooth domains. *Z. Anal. Anwend.*, 9(2):133–153, 1990.

[116] A. H. Schatz and L. B. Wahlbin. Interior maximum norm estimates for finite element methods. *Math. Comp.*, 31:414–442, 1977.

[117] A. H. Schatz and L. B. Wahlbin. Maximum norm estimates in the finite element method on plane polygonal domains. Part 1. *Math. Comp.*, 32(141):73–109, 1978.

[118] A. H. Schatz and L. B. Wahlbin. Maximum norm estimates in the finite element method on plane polygonal domains. Part 2: Refinements. *Math. Comp.*, 33(146):465–492, 1979.

[119] O. Schenk, K. Gärtner, W. Fichtner, and A. Stricker. PARDISO: A high-performance serial and parallel sparse linear solver in semiconductor device simulation. *Future Gener. Comput. Syst.*, 18:355–363, 2001.

[120] L. R. Scott and S. Zhang. Finite element interpolation of non-smooth functions satisfying boundary conditions. *Math. Comp.*, 54:483–493, 1990.

[121] R. Scott. Finite element convergence for singular data. *Numer. Math.*, 21(4):317–327, 1973.

[122] R. Scott. Optimal L^∞ estimates for the finite element method on irregular meshes. *Math. Comp.*, 30:681–697, 1976.

[123] G. Strang and G. Fix. *An analysis of the finite element method.* Prentice–Hall, Englewood Cliffs, NJ, 1973.

[124] R. Temam. *Navier-Stokes equations. Theory and numerical analysis.* North-Holland, Amsterdam, 3rd edition, 1984.

[125] B. Vexler and W. Wollner. Adaptive finite elements for elliptic optimization problems with control constraints. *SIAM J. Control Optim.*, 47(1):509–534, 2008.

[126] D. Werner. *Funktionalanalysis.* Springer, 2008.

[127] G. Winkler. *Control constrained optimal control problems in non-convex three dimensional polyhedral domains.* PhD thesis, TU Chemnitz, 2008. `http://archiv.tu-chemnitz.de/pub/2008/0062`.

[128] J. Wloka. *Partielle Differentialgleichungen.* B. G. Teubner, 1982.

[129] J. Xu. *Theory of Multilevel Methods.* PhD thesis, Cornell University, 1989.

[130] V. Zaionchkovskii and V. A. Solonnikov. Neumann problem for second-order elliptic equations in domains with edges on the boundary. *J. Math. Sci.*, 27(2):2561–2586, 1984.